Understanding
Weather

A Visual Approach

Julian **Mayes** Karel **Hughes**

A member of the Hodder Headline Group
LONDON

Distributed in the United States of America by Oxford
University Press Inc., New York

ARNOLD

First published in Great Britain in 2004 by
Arnold, a member of the Hodder Headline Group,
338 Euston Road, London NW1 3BH

http://www.arnoldpublishers.com
Distributed in the United States of America by
Oxford University Press Inc.
198 Madison Avenue, New York, NY 10016

© 2004 Julian Mayes and Karel Hughes

British Library Cataloguing in Publication Data
A catalogue record for this book is available from the British Library

Library of Congress Cataloging-in-Publication Data
A catalog record for this book is available from the Library of Congress

ISBN 0 340 80611 7

1 2 3 4 5 6 7 8 9 10

Typeset in 10/13 pt GillSans Light by Charon Tec Pvt. Ltd, Chennai, India
Printed and bound in Malta.

What do you think about this book? Or any other Arnold title? Please
send your comments to feedback.arnold@hodder.co.uk

Front cover: NOAA AVHRR false colour composite image of the 'Burns' Day Storm' depression,
13:24h 25 January 1990 (see p. 127).

Understanding Weather

A Visual Approach

Dedicated to the memory of Professor Vince Gardiner,
Head of the Department of Environmental and Geographical Studies,
University of Surrey Roehampton, 1993–98 – an
exceptional and inspiring colleague

Contents

List of boxes

Foreword

The atmosphere can be thought of as a stage that presents a wide variety of performances, some dramatic, others less so. In this book we aim to show how it is possible to understand weather and climate by careful observation of these visual performances. Part of the fascination of the weather lies in the uniqueness of each day's events, but most of these meteorological performances obey general 'rules' that govern the behaviour of the atmosphere.

As in a theatre, we can view the performance from a variety of locations and this visual approach can identify the forces that generate weather and climate. This is not a mathematical or even meteorological text book. Emphasis is placed on careful interpretation of what we can see: from the surface of the Earth we can watch weather, sky and cloudscapes; from meteorological satellites orbiting above Earth we can observe the changing pattern of wind, temperature and moisture as expressed in dynamic cloudscapes. It is this combination of perspectives from below and above that provides us with a new view of the weather, a fusion of meteorology and Earth Observation science.

The structure of the book reflects the integration of these different approaches. The first two chapters of the book focus on understanding the behaviour of the atmosphere, how it responds to variations in heat and moisture (Chapter 1) and how weather systems develop (Chapter 2). The following chapters focus on observation – of weather and cloudscapes from the ground (Chapter 3) and from space (Chapter 4). The final chapters apply this information to examining patterns of weather and climate in two contrasting climatic environments – the mid-latitude climates of Europe and North America (Chapter 5) and tropical climates (Chapter 6).

HOW TO USE THIS BOOK

The aim of this book is to provide a visual approach to understanding weather. Each chapter features a range of case studies and most of these are illustrated with a satellite image and a synoptic chart of the same scene. A brief introduction to interpreting satellite imagery is given in Chapter 1 and this is explored in greater depth in Chapter 4. Most of the weather maps used in the text are surface synoptic charts that show the distribution of air pressure and weather systems at the surface – these are explained in Chapter 2. While a great deal can be learnt from studying satellite images and synoptic charts, your own observation of the sky is undoubtedly a vital first step in understanding weather. We hope that by using this book – and through the references listed at the end of each chapter – you will be able to read the sky and translate what you see and experience into a meaningful understanding of the weather.

The Internet is an invaluable source of current (near real-time) weather information and satellite imagery. It is also an excellent source of colour images and animated sequences. We have resisted the temptation to list many Website addresses because this information often becomes out of date. However, Internet searches using the keywords highlighted in the text in bold will quickly direct you to relevant information. In order to narrow-down your 'hits', we suggest that a specific keyword is preceded by a word that defines the context: e.g., if you want to search for **cold front**, enter <*synoptic climatology, cold front*>; for **reflection**, enter <*remote sensing, reflection*>. Well-defined searches will provide you with a range of definitions that can be more informative than a traditional glossary of terms.

The following Websites provide key information sources:

- the NERC Satellite Receiving Station at the University of Dundee: http://www.sat.dundee.ac.uk – the main UK archive of polar-orbiting and geostationary satellite imagery;
- the UK Meteorological Office: http://www.metoffice.com/ – a wide range of satellite images, synoptic charts and a rainfall radar map;
- EUMETSAT (the European organization for the exploitation of meteorological satellites): http://www.eumetsat.de/en/ – provides access to Meteosat imagery;
- the Royal Meteorological Society, UK: http://www.rmets.org – the society publishes the monthly magazine *Weather*, which includes a Weather Image feature in most issues, usually an interpreted satellite image.

Julian Mayes, Karel Hughes,
Roehampton, London
November 2003

Acknowledgements

We would like to thank a wide range of people who have assisted (sometimes unknowingly) with the preparation of this book.

A wide variety of editorial staff at Hodder Arnold have guided the evolution of this book: Luciana O'Flaherty, Liz Gooster, Lesley Riddle, Abigail Woodman and Colin Goodlad. We thank them for their patience!

We owe special thanks to two staff in the School of Life and Sport Sciences at the University of Surrey Roehampton. It would not have been possible to produce this book without Mary Mackenzie's expertise in drawing almost all of the line diagrams and maps. We would like to thank Mary for taking on this considerable workload with her customary good humour, dealing calmly with our increasing demands. Lionel Gunn provided wise words and invaluable and patient help with computer-related issues that assisted greatly with the compilation of the book. We also thank Debbie Curtis for her timely assistance with several of the figures.

The following organizations and individuals are thanked for allowing us to access substantial sources upon which this book depends:

- the NERC Satellite Receiving Station at the University of Dundee and Neil Lonie for permission to use NOAA AVHRR images which have been incorporated into Figures 1.7(a), 1.8(a), 2.9(a), 2.10(a), 2.11(a), 2.16, 2.17(a), 2.18(a), 3.7, 4.2, 5.4(a), 5.6(a), 5.8(a), 5.9, 5.12, 5.14, 5.15(a), 5.16, 5.17(a), 5.18(a) and 5.19. EUMETSAT is also acknowledged as the source of Meteosat images (obtained via Dundee) used in Figures 1.3, 1.6(a) and (b), 2.14(a–d), 4.10 and 6.3;
- the UK Meteorological Office for permission to redraw sections of synoptic charts originally published in the *Daily Weather Summary* used in Figures 1.7(b), 1.8(b), 2.3(a) and (b), 2.9(b), 2.10(b), 2.11(b), 2.17(b), 2.18(b), 5.5(b), 5.6(b), 5.8(b), 5.15(b), 5.17(c) and 5.18(b);
- Hong Kong Observatory and K. Yeung for allowing us to use up-to-date, precise climatic averages in climate graphs shown in Figures 5.2, 5.20, 5.26, 6.8 and 6.10. These data, supplied by the national meteorological agencies, are available for hundreds of sites around the world on the Website of the World Weather Information Service (http://www.worldweather.org/), maintained by Hong Kong Observatory.

The authors and publishers would like to thank the following for permission to include copyright material in this book:

- Routledge for Figures 1.1, 2.6 and 3.10;
- the American Meteorological Society for Figure 1.2, originally published in Kiehl, J.T. and Trenberth, K.E. 1997: Earth's annual global mean energy budget. *Bulletin of the American Meteorological Society* 78 (2), 197;
- Pearson Education Ltd for Figures 1.4, 2.4 and 5.21;
- University of Wisconsin Madison Space Science & Engineering Centre for Figure 2.5;

Acknowledgements

- Bernard Burton for permission to use a satellite image (shown in Figure 3.19) supplied from his Website: http://www.btinternet.com/~wokingham.weather/wwp.html;
- NASA (National Aeronautics and Space Administration) for Figures 4.1, 4.7, 4.8, 4.9, 4.12, 4.14, 6.1, 6.4, 6.6, 6.9 and 6.11;
- NASA Goddard Space Flight Centre Scientific Visualisation Studio & M Moois Rapid Response Team for Figure 5.25;
- Prentice-Hall Inc. for Figure 6.2;
- NOAA (National Oceanic and Atmospheric Administration) for Figures 6.5 and 6.7.

All of the photographs reproduced in this book were taken by Julian Mayes, with the exception of Figure 3.6, which was supplied by Mr Jim Galvin of the UK Met. Office, who retains copyright.

1

Heat and moisture in the atmosphere

Weather is about energy. Heating from the Sun is the fundamental energy input into the atmosphere and this provides the impetus for the creation of air motion. This energy combines with the plentiful moisture in the Earth's atmosphere to create weather systems. Understanding the behaviour and distribution of heat and moisture in the atmosphere is the key to understanding weather systems and can be achieved by coupling ground-based and space-based (satellite) observations.

1.1 UNDERSTANDING THE ATMOSPHERE: A VISUAL APPROACH

The changing behaviour, movement and moisture of air around us is influenced by the characteristics of the atmosphere above us, in addition to the characteristics of the land and sea around us. Weather is an expression of the movement of air and moisture in three dimensions yet our use of weather maps, and most satellite images, encourages us to think of the atmosphere in two dimensions. In order to understand most weather processes we also need to consider the vertical dimension.

The science of meteorology may seem remote to our experience of weather on the ground. However, understanding the atmosphere above us is easier than it may at first appear because:

- the behaviour of the atmosphere follows certain basic 'rules'. Responses to stimuli such as heat and moisture changes can thus be predicted, although precise outcomes from even the most sophisticated models are unrealistic because the atmosphere displays a significant degree of random ('chaotic') behaviour;
- we can understand changes in the atmosphere by watching cloudscapes, either directly or by means of satellite imagery. The changing form and amount of cloud provides a 'language' that can be read by an observer in order to understand weather.

DEFINING THE SCIENCES OF METEOROLOGY, CLIMATOLOGY AND WEATHER
Meteorology can be described as the science of the atmosphere. Changes in patterns of heat, moisture and motion in three-dimensional space can be described and accounted for by physical laws. However, the basic principles can often be expressed in simple (non-mathematical) terms because most advances in atmospheric science are based on careful observation. This is the approach taken here and the first two chapters provide this visual perspective on understanding meteorological processes.

Weather is the state of the atmosphere as experienced at a given time in a single location, usually on the surface of the Earth. It is the product of meteorological processes acting at any one time and includes such variables as temperature, rainfall, wind speed and the type and amount of cloud. As with meteorology, careful visual observation is the key to identifying and understanding weather situations and this is the main theme of Chapters 3 and 4.

Climate is the state of the atmosphere expressed over a longer time period, comprising both the averages and extremes of weather. Until about the mid-twentieth century climate was regarded as being sufficiently constant to encourage the use of 30-year averages to represent the longer term conditions. The variation of climatic averages over time is now widely acknowledged and this can provide a useful perspective on climatic change. The focus of Chapters 5 and 6 is, respectively, the climatic patterns of the mid-latitudes and the tropics.

OBSERVATIONS FROM ABOVE: SATELLITE REMOTE SENSING

The Earth is now regularly observed from a plethora of space-based platforms; from space we can gain a unique perspective on the weather. Earth Observation science is about remote sensing from above and observing objects of interest from a great distance. By using advanced space and computer-based technologies we are able to gain unprecedented insights into how our planet works.

Meteorological satellites, essentially platforms that carry a payload of instruments, are broadly grouped into two 'families' depending on their orbital path and altitude. They circumnavigate the Earth either by passing over the polar regions (i.e., near-polar orbits) at relatively low altitudes (e.g., 870 km) or are positioned directly above the equator in geostationary orbits at very high altitudes (e.g., 36 000 km) (Box 1.1).

Remote sensing instruments carried on board satellite platforms operate in different ways depending on whether they are 'passive' or 'active' systems. This book focuses on weather images constructed from

---

Box 1.1 Weather satellites and sensors

Most of the images used in this book originate from Europe's Meteosat series of satellites and the Advanced Very High Resolution Radiometer (AVHRR) carried on the National Oceanic and Atmospheric Administration's (NOAA) Polar Orbiting Environmental Satellite (POES) platforms.

Meteosat platforms are geostationary and geosynchronous: they occupy apparently 'fixed' positions above the equator (from an Earth-bound perspective) because their speed and direction are synchronized with that of the Earth's rotation (NOAA's Geostationary Operational Environmental Satellite (GOES) system is similar). At around 36 000 km, the field-of-view (FOV) allows full Earth-disc imaging although the spatial resolution is degraded away from the equator because of Earth's curvature. This 'fixed' orbit is appropriate for acquiring data for weather observations as the Earth is monitored continually and the large area (synoptic) coverage is ideal for tracking entire weather systems. A more advanced, second generation Meteosat (MSG) was launched in 2002.

The AVHRR is carried in a near-polar, Sun-synchronous orbit circling the Earth approximately 14 times daily, crossing the equator at the same local time so that solar illumination is constant and near-global coverage is achieved every 24 hours.

Space-borne sensors measure only selected radiances that have left the Earth system. This is only possible because the atmosphere is transparent to certain wavelengths of energy represented by the 'atmospheric windows' (see Figure 4.4 and Box 4.5) – but there is one exception. The 'water vapour' channels record emissions in a region of strong absorption (see Figure 2.14(d) and Section 4.5).

---

data acquired from passive systems that depend entirely on solar radiation. Radar systems are 'active' as data is acquired by pulsing 'artificial' energy towards Earth and measuring the back-scattered return – solar energy is not involved.

Satellite images are representations of the real world constructed from measurements of the brightness, or intensity, of electromagnetic energy measured by devices remote from the actual scene. Because it is unnecessary, and undesirable, to observe all wavelengths of energy leaving the atmosphere, only carefully selected clusters of signals are measured. These wavelength bands or channels have been chosen because they reveal useful information about the phenomena of interest demonstrated by the different types of weather images used throughout this book (Box 1.1).

1.2 THE THREE-DIMENSIONAL ATMOSPHERE

DEFINING THE COMPOSITION OF THE ATMOSPHERE

The Earth's atmosphere is a shallow 'envelope' of well-mixed gases that provides an essential shield from harmful components of incoming radiation and helps to sustain a habitable environment for living organisms. Most of the mass of the atmosphere is made up of gases that are thoroughly mixed (up to 80 km) and occur in constant proportions: 78.1% nitrogen (N_2), 20.1% oxygen (O_2) and 0.9% argon (Ar). In addition, a range of gases can have a variable concentration over time but are geographically well mixed (Table 1.1). There is also one highly variable and essential gas – water vapour. This usually makes up about 0–4% of the volume of the total atmosphere. Water vapour – along with those gases denoted by an asterisk (*) in Table 1.1 – constitute the 'greenhouse gases', which together generate the greenhouse effect, defined in Box 1.4.

Aerosols are solid and liquid particles, such as dust, salt and sulphates, suspended in the atmosphere. These constituents originate from natural sources, for example, sulphates from volcanic eruptions, or through human activities such as the burning of sulphur-rich coal. Unlike the fixed and greenhouse gases, aerosols are not always resident in the atmosphere for long – they tend to be washed out in rain after a few days. As a result, they rarely become evenly distributed over the Earth's surface. By changing the way in which light is scattered in the atmosphere, aerosols have an important influence on the climate system, including radiation balance and the sky's appearance (Chapter 4).

DEFINING LAYERS OF THE ATMOSPHERE AND AIR PRESSURE

The gas that has the greatest variability over space and time is water vapour. It is mostly concentrated within 12–15 km of the Earth's surface and it is here that weather systems and clouds are created,

Table 1.1 Average concentration of variable gases in the atmosphere

Gas		Parts per million
Carbon dioxide*	CO_2	370
Methane*	CH_4	1.7
Nitrous oxide*	N_2O	0.3
Ozone*	O_3	0.04
Aerosols	various	0.01
Chlorofluorocarbons*	CFCs	0.0001

Note: *denotes greenhouse gases
Source: After Thompson (1998)

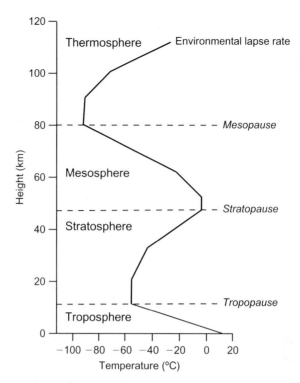

Figure 1.1 *Identifying the main layers of the atmosphere according to the vertical profile of temperature.*

develop and decay – in the region known as the **troposphere**. Understanding weather depends on understanding what is happening in this moisture-rich zone, particularly the process of condensation that gives rise to clouds and rainfall. This is also where the highest density of gases occurs – half the mass of the atmosphere is found in the lowest 5 km.

Density of air affects **air pressure**. The total air pressure exerted by all the gases acting under gravity on the Earth's surface (at sea level) averages around 1013 millibars. One millibar – the standard unit of air pressure measurement – is equal to a force of 100 Newtons m^{-2}. This is sufficient to support about 760 mm or 30 inches of mercury in a tube (a barometer). The average air pressure at around 5 km above ground level is close to 500 mbar. While air pressure measured on the surface changes significantly over time and space, variations measured above 5 km are smaller.

The most important change in the column of air above any location is not that of air pressure but that of temperature. Temperature can either increase or decrease with height and it is upon this distinction that the layers of the atmosphere are defined (Figure 1.1). The troposphere is the initial layer through which temperature usually decreases with height. Within this layer the interactions between atmospheric heat, moisture and motion create local and global weather conditions. This coherent system is driven by solar energy (Section 1.3).

Above about 12 km the temperature, having progressively dropped, remains fairly constant providing a 'cap' to the troposphere defined as the **tropopause**. Above is the **stratosphere**, a region of stability that is without the extensive cloud cover or water vapour found below. Figure 1.1 illustrates the rise in temperature in the stratosphere resulting from the very high ozone concentrations between 25 and 30 km. Ozone strongly absorbs solar radiation and essentially 'blocks' (attenuates) incoming (and outgoing) radiances. It is also found in the troposphere, but at lower concentrations; this distinction is fundamental to understanding the function and effects of this gas (Box 1.2).

Box 1.2 The roles of ozone in the atmosphere

Ozone is essential for life on Earth. Depletion of ozone in the stratosphere is sometimes falsely thought of as a contributor to global warming. Although ozone is a greenhouse gas (so depletion would actually lead to a slight overall cooling) it is a separate environmental issue.

Stratospheric ozone forms by the interaction of short-wave (ultra-violet, u.v.) radiation with oxygen molecules. Ozone, a naturally occurring gas consisting of three atoms of oxygen, is formed as ultra-violet energy splits oxygen molecules. The resulting single oxygen atoms react with more oxygen to form ozone:

$$O_2 + u.v.\ light \rightarrow O + O$$

$$O_2 + O \rightarrow O_3$$

This 'use' of ultra-violet radiation in the stratosphere usually prevents much of it from reaching the Earth's surface.

Ozone is also formed near the Earth's surface from reactions between sunlight and various pollutants, especially nitrogen oxides from vehicle exhausts. It is easily transported to rural environments and is harmful to plant life and to human respiration. Increasing road transport means that **tropospheric ozone** (a greenhouse gas) is increasing in concentration independently from ozone in the stratosphere.

By the late 1990s, as little as one-third of the natural concentration of ozone was being observed in spring (Smithson et al., 2002). Ozone depletion peaks every spring because the reactions that break ozone down are most efficient at low temperatures in the presence of sunlight. Much of the depletion is due to the release of chlorine atoms into the stratosphere by human activity through the invention of the stable compounds chlorofluorocarbons in the 1920s. Chlorine atoms react with ozone, breaking it down to oxygen and chlorine monoxide:

$$O_3 + Cl \rightarrow O_2 + ClO$$

Production of chlorofluorocarbons is now banned by treaty and the rate of ozone depletion is already slowing (Fischer and Staehelin, 2003). Ozone depletion may be capable of successful long-term management but we remain committed to several decades of heightened ultra-violet radiation and to an increased risk of skin cancer. This illustrates the problems caused by the long **atmospheric lifetime** of gases.

Kinetic energy released in the formation of O_3 leads to warming of the stratosphere around the ozone layer resulting in the rise in temperature shown in Figure 1.1. Above the stratosphere, temperature once again drops with height within the mesosphere. Temperature increases again in the upper thermosphere owing to the absorption of ultra-violet radiation by molecular oxygen.

1.3 ENERGY IN THE ATMOSPHERE

RADIATION FROM THE SUN

The Sun is the primary source of energy in the atmosphere. The surface of the Earth and the atmosphere are heated by solar radiation – this is predominantly short-wave radiation, since

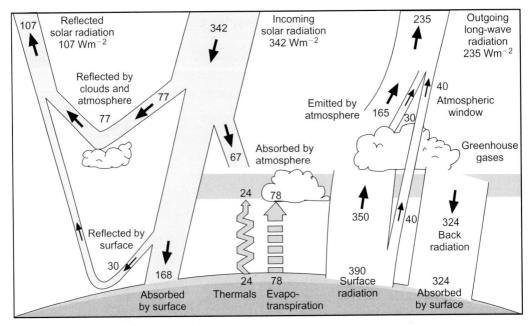

Figure 1.2 *The average distribution of short-wave and long-wave radiation in the atmosphere (from Kiehl and Trenberth, 1997).*

wavelength of radiation emitted from any surface is inversely related to the temperature of the surface (Chapter 4). On reaching the upper boundary of the atmosphere, a significant proportion of solar energy is absorbed by stratospheric ozone (Box 1.2) but most of the energy that is not attenuated reaches the troposphere. As the air is denser at lower levels, considerable interference occurs as the transmitted energy interacts with atmospheric constituents and is absorbed, scattered and reflected (Figure 1.2). These processes mostly affect shorter wavelengths of energy; longer wavelengths tend to have an uninterrupted journey.

Just under 50% of the downwelling radiation actually reaches the Earth's surface either by direct radiation (>27%) or by diffuse radiation (scattering); the rest is reflected, mostly by clouds (around 20%) or is absorbed by the atmosphere. At the Earth's surface, most of this radiation is absorbed and converted into heat energy – especially by water; the rest (<9%) is reflected back into the atmosphere (Figure 1.2).

ALBEDO

The Earth's surface temperature is predominantly determined by the amount of short-wave radiation absorbed. The amount absorbed is proportional to the amount reflected – a ratio expressed by the **albedo** value, the proportion of incoming radiation reflected by a surface (Box 4.4). The albedo of a surface depends on its physical and chemical properties. Highly absorptive surfaces have low albedo values (e.g., water), retain energy and, consequently, warm; surfaces such as cloud-tops, ice and snow have high albedo values, and remain cold because they are strongly reflective. As much of this reflected energy is 'lost' to space, it is no longer available for warming the Earth, hence the cooling effect of clouds.

Global variations in albedo can be identified in visible satellite images because the brightness (whiteness) or dullness (greyness or blackness) of features depends on their reflectivity (Box 1.3; Figure 1.3). Clouds mostly appear bright in visible images because their tops are highly reflective. Below the clouds, the less reflective land surfaces appear dull; the oceans reflect least energy and usually appear black. Albedo of low cloud can be as high as 0.7 to 0.9, though high cloud such as cirrus has a much lower albedo.

Box 1.3 Visible images

Visible images represent a top-down view of the weather that is familiar, but by no means the same as the view from below. Space-based sensors detect energy scattered and reflected from cloud-tops and from the Earth's surface through gaps between the clouds. Clouds are 'frontline' reflectors and appear bright (white to light grey) in images; thicker clouds (e.g., cumulus) at any height, appear brighter than thinner clouds (e.g., cirrus). Earth's surfaces, excluding ice and snow, generally reflect far less energy and appear dull or dark, especially water, which absorbs most incident energy and appears black.

Figure 1.3 is a visible (VIS) image acquired from Meteosat using a very broad wavelength band that detects wavelengths between 0.4 and 1.1 μm and includes reflected visible light energy and reflected near infrared wavelengths (i.e., beyond 0.75 μm) transmitted out to space through an atmospheric window (Figure 4.3, Box 4.5). An advantage of stretching into the near infrared is that atmospheric scattering is reduced as wavelengths lengthen so the outgoing signal is enhanced and images appear brighter.

Figure 1.3 *Meteosat visible image for 12:00 h 23 May 2003. Clouds appear light because they have a high albedo and the ocean appears dark because of the low albedo. Land areas have an intermediate and rather variable tone (according to local variations in albedo; the high albedo of deserts is clearly shown by the Sahara).*

LONG-WAVE RADIATION FROM THE EARTH'S SURFACE

The heating at the surface of the Earth derived from solar radiation is, in turn, lost to the atmosphere as long-wave (outgoing) radiation. Only about 5% of this energy loss passes directly out of the atmosphere to space. Just as solar radiation is absorbed by ozone and various aerosols, so long-wave radiation is absorbed by the greenhouse gases (Box 1.4). Their absorption of energy helps to retain heat energy in the atmosphere; indeed, the resulting greenhouse effect raises surface temperature by about 33–34 deg. C. It can be seen therefore that the greenhouse effect is a natural phenomenon that is vital to life – without this trapped heat the planet would be a very cold place (around −18°C).

The greenhouse effect can be demonstrated by links between water vapour, clouds and temperature; water vapour is the one greenhouse gas that can change its local concentration quickly. This is why cloudy nights are usually milder than clear nights because water vapour in the cloud absorbs and emits long-wave energy that warms the lower troposphere. Hot deserts are often cold at night because there is little water vapour in the atmosphere here to absorb the heat released through terrestrial (long-wave) radiation.

Box 1.4 The greenhouse effect and greenhouse gases

The concentration of greenhouse gases (Table 1.1) has increased as a result of a range of human activities. For example, following the Industrial Revolution, emissions from the widespread use of fossil fuels (oil, coal and gas) have contributed to changing the composition of the atmosphere. Since 1750, the amount of carbon dioxide has increased by 31% and, as this gas plays a critical role in 'trapping' energy, more gas is now present to absorb more energy, thus enhancing the greenhouse effect. Many scientists believe this to be a major cause of global warming (Drake, 2000).

Currently, the Earth's surface temperature is around +15°C. It is predicted to rise by between as little as 1.4 deg. C and as much as 5.8 deg. C by 2100 (Intergovernmental Panel on Climate Change (IPCC), 2001). Scientific uncertainty exists because feedback processes operate. As warmer air holds more water vapour, further warming means that additional water vapour will be available to enhance the greenhouse effect – a positive feedback. However, increasing amounts of thick cloud could have a cooling effect (negative feedback) by blocking incoming radiation and offsetting some of the original warming (but see Box 4.6).

THE BALANCE OF RADIATION ON THE GROUND AND SURFACE TEMPERATURE

The relationship between short-wave and long-wave radiation is the fundamental influence on surface temperature. This can be expressed in terms of **net radiation**:

$$\text{net radiation} = \text{short-wave} - \text{long-wave radiation}$$

Positive values of net radiation indicate that more radiation is being received than lost by the surface at a particular time. The Earth's surface receives far more short-wave radiation than it loses as long-wave. The resulting positive net radiation would imply a warming – in theory – of about 250 deg. C per day! (Thompson, 1998.) This does not occur because the atmosphere itself has a negative radiation balance – heat energy is lost to space. The atmosphere thus controls the transmission of heat from the Earth's surface towards space. This process – in which the greenhouse effect provides a kind of thermostat – provides the atmosphere with energy. It is the driving force for the atmospheric system, including global winds, cloud and rainfall.

The surface and atmosphere together have a net radiation value close to zero – this balance between incoming and outgoing radiation maintains a generally constant temperature in the atmosphere. Of course, this balance is disturbed by any enhancement to the greenhouse effect above the 'natural' pre-Industrial value.

HEATING OF AIR AND CONVECTION

If a surface receives more radiant energy than is 'lost' (positive net radiation) air becomes heated from below. As heat always travels from a hotter to a cooler environment, it is **conducted** (transmitted) away from the ground but only for a short distance (just centimetres). Once the lowest layer of air has been heated, the buoyant, upward motions of **convection** currents distribute the heat more widely, convection being the buoyant, upward motion in liquids or gases that are heated from below.

In the atmosphere, a bubble of warm, buoyant air near the ground is called a 'thermal'. As rising thermals cool, their density increases and the rising current of air will stop when it is no longer more buoyant than its surroundings. As heating continues, so the convection current will strengthen and deepen and thermals will reach a greater height, a useful principle in understanding cloud development. Convection is the main process of heat transfer into the atmosphere during the daytime.

THE INFLUENCE OF LATITUDE AND THE SEASONS

Seasonal changes in temperature are caused by the changing elevation of the Sun above the horizon. The tilt of the Earth's axis of rotation leads to the Sun being overhead at the Tropic of Capricorn on 21 December and over the Tropic of Cancer on 22 June – the two solstices. In southern Britain, the Sun is only 16° above the horizon at midday in December (Smithson et al., 2002). The strength of the low Sun is therefore weakened by the increased distance of the atmosphere through which the radiation passes (increasing scattering) and by the greater surface area intercepted by the solar beam.

At each solstice the Sun shines continuously at the polar regions of one hemisphere. At present, the Sun is actually closest to Earth on 3 January and furthest away on 4 July, because the Earth has an elliptical orbit around the Sun. Although this might be expected to lead to milder winters and cooler summers in the northern hemisphere, this effect is counterbalanced by the greater land-mass in this hemisphere and by differing heat circulation.

The Sun is over the equator at the Equinoxes (21 March and 22 September). Other factors being equal, receipt of solar radiation at these times should be equal in both hemispheres. However, these dates are not exactly 6 months apart – the summer half-year in the northern hemisphere is, at present, 5 days longer than that of the southern hemisphere. Again, this is a function of orbital cycles and is subject to gradual change over many thousands of years (Barry and Chorley, 1998).

Figure 1.4 shows the effect of latitude on average annual net radiation. The cooling effect of negative net radiation at high latitudes is enhanced by the high albedo (reflectivity) of the poles, which occurs as a result of ice and snow cover.

THE INFLUENCE OF TIME OF DAY

Just as the temperature of the seasons is controlled by the Sun's angular relationship with the Earth, daily (diurnal) temperature changes can also be related to the Sun angle since this determines the

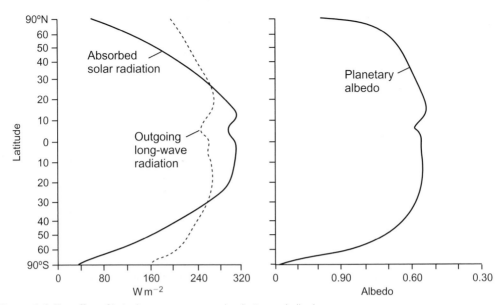

Figure 1.4 *The effect of latitude on average annual radiation and albedo.*

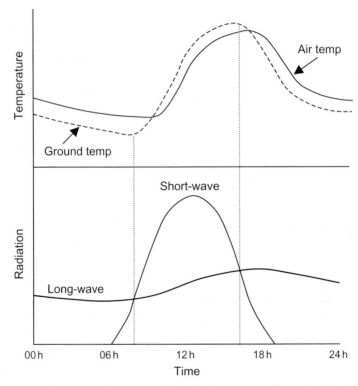

Figure 1.5 *The variation of incoming and outgoing radiation and temperature for a typical cloud-free day.*

local radiation balance. The variation between radiation balance and surface temperature for a typical cloud-free day is shown in Figure 1.5. If skies remain clear of cloud, the amount of incoming radiation reaching the surface is clearly a function of Sun angle. Long-wave radiation is continuously emitted from the Earth's surface but it increases following daytime heating. Net radiation is negative (representing a loss of heat energy) when the outgoing long-wave radiation exceeds the incoming solar energy.

It is usually about half an hour after sunrise that a balance is reached when net radiation is zero – the Sun has to rise above the horizon by a small angle before the solar radiation gained balances the long-wave loss. Similarly, net radiation reaches zero shortly before sunset as the diminishing incoming radiation is temporarily in balance with outgoing radiation. While net radiation influences ground temperature, air temperature will respond to variations in ground temperature through conduction, turbulence and convection.

As distance above ground increases, the diurnal range in temperature and the rate of temperature change both decrease. Temperature extremes are concentrated at the surface that is open to the atmosphere – it can be the surface of tarmac or soil, the top of a crop or the tree canopy. This surface, where energy is absorbed by day and released to the atmosphere at night, is called the **active surface**.

1.4   RATES OF WARMING AT THE SURFACE OF THE EARTH

HEAT CAPACITY

The speed at which a surface can warm is determined by the **heat capacity** (or specific heat) of that surface. It is defined as the amount of energy required to raise the temperature of a given mass by 1 deg. C. Water has a relatively high value ($1.0\,cal\,g^{-1}$ per °C), five times that of dry soil. Water is relatively transparent to solar radiation, meaning that heat energy from the solar radiation is spread widely whereas on land it is concentrated at the surface.

Because air is a poor conductor of heat, the lowest heat capacities occur over soils that contain a large amount of air. The effects of heating by day and cooling by night are then concentrated in a much narrower zone at the surface, with little conduction of heat to lower levels. This is the reason why sandy soils (in which the soil particles are separated by a large amount of air) warm up quickly in spring but are also conducive to severe cooling at night. Air and ground frost is a hazard for much of the year in the mid-latitudes where the soil is sandy. If the surface is dry, the heat capacity is reduced, increasing the rate of cooling at night.

THE INFLUENCE OF LAND AND SEA

The heat capacity of the sea is about ten-times that of the land. This is because the upper layers of water are transparent to solar energy and, as the sea is in constant motion, the heat is distributed across a considerable depth. Up to 20% reaches a depth of 9 m and, in summer, a slight energy receipt may be recorded down to 40 m (Barry and Chorley, 1998). Unlike the marked diurnal change seen on land (often 6–10 deg. C), sea surface temperatures (SST) change little between night and day, varying by just half a degree in shallow, off-shore waters. In deep water temperature changes are negligible.

The thermal state of land and sea surfaces can be detected from space (Box 1.5). Satellite sensors can be 'tuned' to respond to emitted long-wave radiation transmitted up through a cloud-free sky. Figure 1.6 is a thermal infrared image of the same scene shown in the visible version (Figure 1.3). Night and day scenes can be recorded because thermal infrared sensors work independently of sunlight. Images reveal brightness temperatures, not the actual temperature of surfaces (see Box 4.3). Infrared images demonstrate clearly contrasts in day and night temperatures. The scenes across northwest Europe (Figures 1.7 and 1.8) reveal these contrasts and highlight the daytime heating of land, especially on relatively cloud-free days.

1.5 EFFECTS OF TEMPERATURE CHANGES – STABILITY AND INSTABILITY

Having reviewed the factors that influence heating of different parts of the Earth's surface, it is now appropriate to consider the effect of these temperature changes on the air above. Temperature is an important influence on cloud patterns, location and characteristics.

STABLE AND UNSTABLE AIR

Air temperature has a variable and irregular pattern of change with height, unlike air pressure that decreases more steadily. Temperature variations lead to fundamentally different states in the atmosphere: stability and instability (Box 1.6).

Detailed explanation of these concepts can sometimes appear rather complicated but it is possible to gain a good understanding from a non-mathematical, step-by-step approach to account for changes in the sky that can be observed by the keen sky-watcher.

Figure 1.9 shows typical cloud types associated with stable and unstable air. While stable air encourages cloud to spread horizontally in a sheet-like form, unstable air encourages vertical growth resulting in a more broken cloud distribution. Note that it is not accurate to associate stable air with a lack of cloud, or a lack of activity in the troposphere.

All clouds are formed from the condensation of water vapour into cloud droplets (or the sublimation of supercooled vapour to ice crystals). The source of the water vapour is the evaporation of water from the surface of the Earth or by the transpiration of moisture from plants.

Box 1.5 Thermal infrared images

These images make the invisible visible because they represent heat escaping into space (e.g., Figure 1.6) Sensors record the intensity (brightness) of long-wave infrared energy emitted from the Earth–atmosphere system. The images often appear 'brighter' than visible images of the same scene because longer wavelengths experience little scattering (less interference) in the atmosphere. In night images, high, cold cloud-tops appear very bright, cool land below appears dull and relatively warmer oceans appear black. In daytime images, warmer land appears dull relative to the cooler (brighter) oceans (Figure 1.6). The grey-tone coding is explained in Chapter 4 (Box 4.1).

Meteosat employs two thermal channels (2 and 3) that detect infrared radiances escaping to space (Figure 4.3). Channel 3 measures emissions at wavelengths between 10.2 and 12.5 μm that are transmitted through an important atmospheric window; Channel 2 records shorter emitted infrared wavelengths peaking around 6.7 μm: this is the water vapour channel actually located within an atmospheric absorption zone (Figure 1.6(b)). NOAA's AVHRR system (e.g., Figure 1.7(a), Figure 1.8(a)) employs two channels to detect these emissions (equivalent to Meteosat's broad-band Channel 3) between wavelengths 10.3 and 11.3 μm (Channel 4) and 11.5 and 12.5 μm (Channel 5).

(a)

(b)

Figure 1.6 *Meteosat imagery (12:00 h 23 September 2003): (a) thermal infrared image; (b) water-vapour image. Image (b) demonstrates how much water vapour exists in the atmosphere. Bright areas are moisture-rich, dark areas are water vapour 'holes' indicating very dry air. Clouds can be identified in both images.*

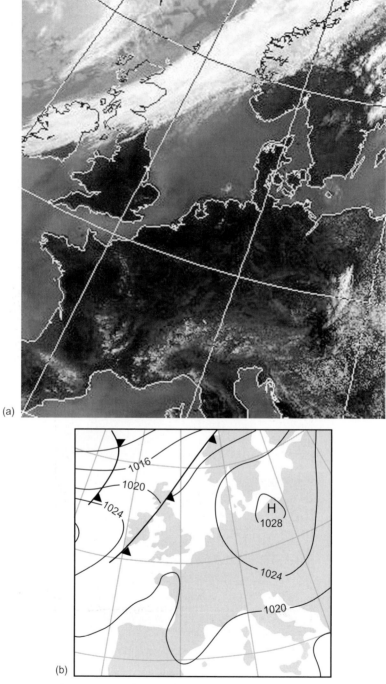

(a)

(b)

Figure 1.7 *Northwest Europe on a hot day: (a) daytime surface temperature as inferred by NOAA AVHRR thermal infrared imagery; 12:37h 3 August 1990; (b) surface synoptic chart for 12:00h. Until 2003, this was the day of the highest temperature recorded in the British Isles, 37.1°C at Cheltenham, Gloucestershire. The contrast between the warmth of the land areas (black) and the cold cloud-tops (white) on the cold front approaching Scotland is clearly shown. The sensor measures only thermal emissions from the highest cloud layer: since cloud-top temperatures are inversely related to height, higher, colder clouds appear brighter (lighter tone) than warmer, lower clouds (see Chapter 4). The sea surface appears colder (lighter) than the land as this was recorded at midday.*

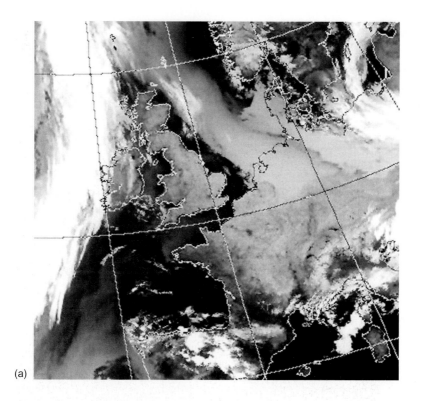

(a)

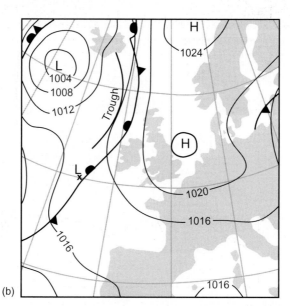

(b)

Figure 1.8 *Northwest Europe at night: (a) NOAA AVHRR thermal infrared image for 02:40 h 25 July 1990;*
(b) surface synoptic chart for 12:00 h 25 July 1990. This satellite image shows grey clouds over the North Sea indicating
relatively warm, and hence relatively low altitude, cloud-tops. The cloud to the west of Britain is associated with deeper
cloud on a cold front with lower cloud-top temperatures. Where skies are clear the sensor is able to detect differences
in surface temperature; being a night-time image, the land is cooler than the sea (although the urban heat-islands of
Paris and London appear dark).

Box 1.6 Definitions of stable and unstable air

Stable air exists when a parcel of air is less buoyant than the surrounding air; vertical movement is discouraged unless the air is forced to rise (absolute stability).

Unstable air exists when a parcel of air is more buoyant than the surrounding air – including the air above (absolute instability); the 'lighter' air will tend to rise until a height is reached where all the air has the same degree of buoyancy.

The buoyancy of air is controlled by temperature, hence the importance of temperature variations through the atmosphere.

(a)

(b)

Figure 1.9 *Cloudscapes of (a) stable air (stratocumulus); (b) unstable air (cumulus).*

CONDENSATION

Condensation occurs when the amount of water vapour exceeds the capacity of air to hold it; this may happen simply with the addition of extra water vapour if, for example, air passes over a large body of water. The extent to which air can hold water vapour is related to its temperature. For any particular temperature, there is an upper limit beyond which further water vapour cannot be held – this limit decreases as air is cooled. Any further addition of vapour results in the air becoming saturated and condensation of cloud droplets taking place. The temperature at which a sample of air becomes saturated is called the **dewpoint temperature**. This is a critical variable in weather forecasting because it indicates the **temperature** at which cloud formation will start.

In order to calculate the dewpoint temperature we need to express the actual measured amount of water vapour in the air as a proportion of the maximum amount that the air can hold at that temperature. This value, expressed as a percentage, is **relative humidity** (see Chapter 2).

TEMPERATURE FLUCTUATIONS, CONDENSATION AND CLOUD FORMATION

Condensation and cloud formation respond closely to changes in air temperature. If you were to rise a few hundred metres through the air with a thermometer, the temperature profile could be recorded (Figure 1.10). This is the **environmental lapse rate** – the cooling (lapse) rate of the air. Any warming with height is referred to as a negative lapse rate.

For example, assuming a sunny cloud-free day, the morning Sun heats the ground, promoting a certain rate of cooling (lapse rate) with height. Because this warming weakens with height, the environmental lapse rate (cooling rate) at midday will be greater than that observed earlier in the day – the ground will have warmed more than the air some kilometres above.

What will be the effect of having created this layer of warm air close to the surface? Since the density of gases is inversely related to temperature, this layer will become buoyant and will try to rise. Bubbles of air will start to lift off the ground – thermals. As they rise they will be forced to expand because they are passing into a layer of lower atmospheric pressure. A balloon released into the open air demonstrates this: the air inside expands as it drifts upwards through the low pressure atmosphere – and then bang!

Reduced air pressure forces the air molecules within the bubble to work (use energy) to occupy a greater volume of air. The bubble of air will cool progressively as it rises. This is called **adiabatic cooling** (arising from adiabatic expansion); in other words, the cooling and expansion occur without any direct input of energy into or out of the bubble – the change was an indirect consequence of the energy used up in being forced to expand. This cooling is related to the distance the bubble rises. It can therefore be expressed as the **dry adiabatic lapse rate** (see Box 1.7), the temperature of a moving bubble of air (not the surroundings). A moving bubble of air can therefore have an independent temperature to its surroundings, a vital distinction in meteorology. The greater the warmth of the bubble compared with its environment determines the buoyancy of the bubble – this is termed the **Convective Available Potential Energy** (CAPE).

The dry adiabatic lapse rate (DALR) is a constant – rising unsaturated air bubbles will cool at a rate of 0.98°C per 100 m, as shown in Figure 1.10. Clouds will only develop if rising bubbles of air cool sufficiently to become saturated and this will depend on how high they rise. This is determined by the density of the bubble in comparison with its surroundings at the same height and can easily be determined by comparing temperatures. Bubbles of warm air tend to rise, and continue to rise, because they are less dense and more buoyant than the surrounding air that cools at a much faster rate: in these circumstances the air is described as unstable.

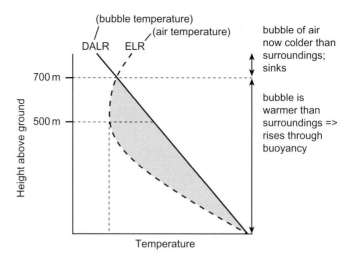

Figure 1.10 *The variation of air temperature measured above the Earth's surface (the environmental lapse rate (ELR)). The temperature of a small bubble of air rising through the atmosphere is shown by the dry adiabatic lapse rate (DALR). The shaded area between the line and the curve represents the temperature difference between the bubble and its surroundings, a measure of the energy available for convective uplift – convective available potential energy (CAPE).*

In the example given (Figure 1.10), the environment begins to get warmer at around 500 m. The bubble becomes cooler than its surroundings above 700 m and, as it is now denser and less buoyant, it will tend to descend. This 'capping' process, by the layer of stable air, inhibits any further upward movement of the thermal. This situation results in a temperature inversion; this is when air temperature decreases with height (above 500 m in Figure 1.10).

CONDENSATION OF RISING AIR

Air always contains water vapour and can become saturated either by cooling or by the addition of more water vapour. Over the sea, air can gain moisture as seawater evaporates into it. Over land it is more commonly saturated by cooling, often associated with the rising bubbles of air in an unstable environment. If the dewpoint of air shown in Figure 1.11 is 10°C, rising thermals are cooled to this dewpoint temperature at 400 m (shown by the DALR line). Small shreds of cloud will soon start to condense here. The bubbles of rising air will still be buoyant and unstable and will thus continue to rise.

Condensation involves a change of state from gas to liquid. This results in energy being released into the surroundings as the molecules are rearranged. This energy release is called **latent heat**. Condensation leads to a release of latent heat, while the opposite change of state – evaporation – results in energy being absorbed from the surroundings, creating a cooling effect. The warming associated with condensation counteracts some of the original cooling accompanying dry adiabatic uplift: the bubble is no longer cooling at the DALR (0.98°C per 100 m) but at the saturated adiabatic lapse rate (SALR) of around 0.65°C per 100 m. The SALR depends on the amount of water vapour available for condensation. As warmer air releases more latent heat into the atmosphere, the SALR is lower at higher temperatures (represented by the slight curve in Figure 1.11).

Condensation has an effect on air stability – it increases instability! Once condensation occurs (above 400 m in Figure 1.11), the bubble of air becomes warmer and remains unstable and buoyant with enough energy to rise even higher in the atmosphere, spreading cloud development up to 1000 m. Condensation is therefore an important source of heat energy in the global climate, especially

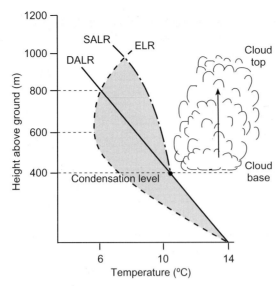

Figure 1.11 *The effect of saturation on the temperature of a rising bubble of air – the saturated adiabatic lapse rate (SALR).*

in the tropics where the SALR is relatively low. The vast amounts of condensation in tropical cyclones, for example, provide spectacular injections of energy into the weather systems.

To summarize these interrelated processes, the main concepts encountered in studying temperature changes in the atmosphere are reviewed in Box 1.7. The distinction between stable and unstable air is crucial in distinguishing different cloud and weather types.

VARIATIONS OF INSTABILITY OVER TIME AND SPACE

Figure 1.12 shows how environmental lapse rates typically change on a sunny day in response to surface heating and how the stability of air has changed over this time.

The 07:00 h ELR temperature curve in Figure 1.12 shows a typical **surface inversion** of temperature typical of a clear night. Nocturnal cooling begins at ground level, because of terrestrial radiation, and later spreads upwards into lower layers of air. Now the ground is relatively cool, air will not rise freely because it will have a greater density than the air above, creating a layer of stable air. By 13:00 h the surface temperature has reached 20°C, but the air above will have warmed much less, owing to the slow rate of heat transfer through air. As a result, the temperature difference with height (environmental lapse rate) will have increased and the air will now be unstable.

The ELR often lies between the SALR and the DALR; i.e., the environment cools at between 0.6°C and 1.0°C per 100 m. The air will then be stable if it is unsaturated (bubble would be cooling at 1°C per 100 m and thus become cooler than the surroundings) but unstable if it was saturated (bubble cooling at only 0.6°C per 100 m and thus become warmer). The instability is then said to be conditional upon the moisture content of the air – this is **conditional instability**.

Conditional instability is common in the mid-latitudes. In the early part of the ascent of an air bubble, where the air is unsaturated, free convection will not be triggered because the air will be stable (Figure 1.13). Air bubbles are often forced to rise through this stable layer because of passage over high ground or by uplift at fronts. In these situations, rising air bubbles may be lifted and cooled to dewpoint temperature. The release of latent heat and consequent cooling at the SALR may enable the

---

Box 1.7 A summary of lapse rates and instability

The changing temperature of air above the ground is a fundamental variable determining condensation and cloud formation. This can be identified according to condensation and lapse rates:

Environmental lapse rate (ELR)	The cooling rate of the air in general above one point. Average value 0.65°C per 100 m.
Dry adiabatic lapse rate (DALR)	The cooling rate of a bubble of air rising through the atmosphere. A constant value: 0.98°C per 100 m.
Dewpoint temperature	The temperature below which air will become saturated, resulting in condensation.
Saturated adiabatic lapse rate (SALR)	The cooling rate of a bubble of rising air that has been cooled to (or below) its dewpoint temperature. The release of latent heat offsets some of the original cooling – average value 0.6°C per 100 m.

The combination of these lapse rates gives us the temperature of any rising bubble of air and that of its surroundings. If it is warmer, it will continue to rise; if it is cooler, it will descend.

Unstable air	The upward motion of thermals is encouraged because the surroundings are cooler than the rising bubbles of air.
Stable air	Any upward motion (apart from any forced uplift) is discouraged because the ELR is too small to allow thermals to rise (<0.98°C per 100 m in unsaturated air; <0.65°C per 100 m in saturated air). Cloud may develop but will spread horizontally rather than vertically (though some forced uplift often occurs).
Temperature inversion	A rise in air temperature with height (ELR is usually considered to be negative to represent this warming with height). Leads to very stable air being created. Can be observed either at the surface (such as on a clear night) or higher up (where there is widespread descent of air).

---

bubble of air to become unstable if it now becomes warmer than its surroundings: in Figure 1.13 this is achieved at 500 m and convective cloud will form above this level.

Convective instability occurs when a deep layer of rising air is moist at its base but drier towards the top. As the moist air is likely to become saturated before the drier air above, it will cool at a slower rate, producing an overall increase (steepening) of the ELR.

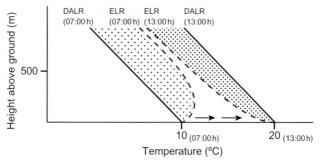

Figure 1.12 *Variations in environmental lapse rates over a cloud-free day showing the effect of changing surface temperature on the stability of the air above.*

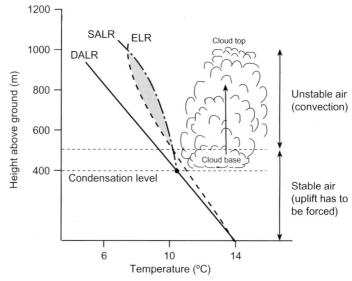

Figure 1.13 *Conditional instability – the air is stable up to 500 m. Between 400 and 500 m any rising bubble of air is stable but saturated and cools at the SALR and a layer cloud will form. Between 500 and 1000 m the air is unstable, being warmer than the surroundings; the grey shading represents the convective available potential energy.*

1.6 FROM CLOUD DROPLETS TO RAINDROPS

The change from cloud droplets to raindrops depends on the condensation of water vapour and is associated with a dramatic increase in droplet size: with diameters around 1–2 mm, most raindrops are over 100 times larger than cloud droplets. Up until the early twentieth century, it was thought that cloud droplets grew until they fell to Earth as raindrops. New theories of raindrop formation, confirmed by observation, have replaced this notion and there is now an improved understanding of the initial condensation process. Condensation rarely takes place spontaneously at the dewpoint temperature – **condensation nuclei** are also needed.

Condensation nuclei are suspended particles (aerosols) offering surfaces upon which the process of condensation occurs. Their distribution is highly variable; high concentrations occur in stable air with light winds; locally, they are washed out by rain. The particles originate from natural sources (e.g., dust, sea salt) and human activities (e.g., liquid and solid pollutants). The idea that condensation occurs against surfaces rather than in free, pure air can be seen when dew forms on blades of grass after a cold night; other surfaces such as spiders' webs may also have water droplets on them.

THE GROWTH OF RAINDROPS

There are two widely accepted theories to account for the rapid development of raindrops. The most widely applicable is the **Bergeron–Findeisen** process, suggested by the Norwegian meteorologist Bergeron in 1933, and confirmed by the German, Findeisen, a few years later.

Across much of the world clouds grow sufficiently high to enter areas where the temperature is below the freezing point of water – in other words, they rise above the 'freezing level'. Bergeron realized that these clouds would contain a mixture of water droplets and ice crystals. Since water does not instantly freeze at 0°C, it may remain in liquid form even when the surroundings are well below freezing (i.e., in a 'supercooled' state). Ice crystals and supercooled water droplets often coexist, but because they respond in different ways to the saturation process, air that is saturated with respect to water will be supersaturated with respect to ice. Saturation occurs more readily around ice crystals and it follows that water droplets then either evaporate or freeze onto the surfaces of adjacent ice crystals. The end result is that ice crystals grow at the expense of water droplets. As ice crystals grow, they start to fall and start to 'zap' adjacent water droplets and ice crystals, leading to further growth and faster descent. As the crystals grow they develop into snowflakes.

The combination of ice crystals and water droplets occurs frequently in clouds in many parts of the world. Falling ice crystals develop into snowflakes as they fall. These melt into raindrops if the surface temperature is more than about +4°C. Clouds above the mid-latitudes commonly have temperatures between −10 and −35°C and it is within this range that the combination of water droplets and ice crystals is conducive to the Bergeron–Findeisen process.

Langmuir's chain reaction theory explains rainfall in the absence of any freezing process. It was noted in the mid-twentieth century that rain in the tropics fell from warm, ice-free clouds because they were not high enough to reach the freezing level: an explanation other than the Bergeron–Findeisen process was needed, and this involves the **collision and coalescence process**. Water droplets collide and coalesce ('bump' and 'clump' together) as a result of atmospheric turbulence and convection. A 'chain reaction' occurs because, as these larger droplets reach a critical radius of 3 mm, they become unstable and break up, forming lots of smaller droplets that continue to collide and thus grow to large droplets by this chain reaction process.

1.7 FORMS OF PRECIPITATION

Precipitation is the name given to all **forms** of water and ice reaching the Earth's surface from the atmosphere either by falling from clouds or by direct deposition (Table 1.2).

Different precipitation **types** occur because of different methods of uplift in the atmosphere (Table 1.3). Precipitation can develop in (thermally) **stable** air by **forced uplift**. This can be due to either convergence or orographic uplift over high ground, uplift that is typically slow but widespread. **Free uplift** occurs in **unstable** air – this can be much faster but is usually less persistent than forced uplift. Precipitation resulting from convectional uplift in unstable air is a **shower** (a brief spell of precipitation arising from non-cumulus clouds – i.e., in stable air – is termed 'intermittent precipitation' if it lasts for less than one hour and should not be referred to as a shower).

Table 1.2 Forms of precipitation and their sources

Type of precipitation	Source
Rainfall	Water droplets formed from the growth of cloud droplets
Snow	Ice crystals formed from the aggregation of ice crystals in cloud
Sleet	A mixture of rain and snow, often partly melted snowflakes
Hail	Solid usually spherical lumps of ice produced in unstable air; usually composed of concentric shells of ice deposited after repeated uplift through ice crystals in cumuliform cloud. Often occurs when surface temperatures are high
Dew	Direct deposition of water droplets on the ground surface due to condensation

Table 1.3 A summary of different types of precipitation according to the method of uplift

Method of uplift	Cause	Speed
Orographic uplift	Passage of air over high ground	Variable
Convergence (frontal) uplift	Large scale horizontal convergence of air between adjacent air masses	Gradual
Convective uplift	Free uplift of buoyant air when the atmosphere is unstable	Potentially fast

1.8 THE FÖHN EFFECT – A LOCAL INFLUENCE UPON CLOUD AND RAINFALL

The föhn effect is the name given to a warm, often dry wind that descends from high ground. It is an important influence on the distribution of cloud, rainfall and temperature around hills and mountains. 'Föhn' is the general name of the phenomenon but it is known by a wide array of local names, the most well known being the *Chinook* of North America.

When air descends, it warms because of adiabatic compression (the reverse of adiabatic expansion) as the larger, denser air mass above 'presses' it down. When this happens over land, the warming air gains little moisture from the underlying surface and remains, or quickly becomes, unsaturated. The warming rate of unsaturated air is given by the dry adiabatic lapse rate of 1°C per 100 m (that is, a warming at this rate while the air descends). As air continues to descend, it dries out as its moisture-holding capacity increases, giving rise to the **rain-shadow** phenomenon as the higher temperatures ensure that cloud droplets evaporate, often leaving a clear sky (Figure 1.14).

The föhn effect is most typically associated with situations when air becomes saturated ascending windward slopes, leading to a reduction in the cooling rate from the DALR to the SALR (0.65°C per 100 m). Once over the hill, air will mostly warm up at the faster DALR, as explained above, and the result is a higher temperature on the lee side of the high ground (Figure 1.15).

The föhn effect also occurs in fine weather in the absence of saturation. When air is very stable in high pressure situations, low-level air may be blocked by hills. Consequently, only air that originates from summit height reaches the lee-side and is therefore only subject to adiabatic warming.

Temperatures can rise by as much as 20 deg. C to the lee of major mountain ranges where the föhn wind can be strong and gusty, such as over the Rocky Mountains of North America (Quaile, 2001) where the wind is called a *Chinook* (literally, 'snow-eater' in native American Indian language). Changes in temperature and moisture can be felt over 100 km downwind.

Figure 1.14 *Breaks in the cloud to the lee of Exmoor in southwest England as a föhn wind descends into a rain-shadow area. The photograph was taken looking west, towards the high ground; altostratus over the hills (in the centre of the photograph) is breaking up to reveal blue sky with a little cirrus overhead.*

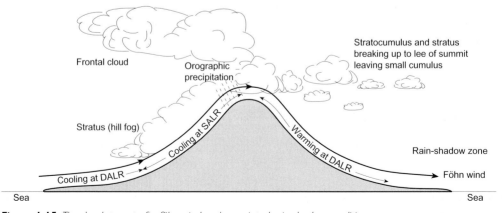

Figure 1.15 *The development of a föhn wind and associated rain-shadow conditions.*

Wherever the föhn effect is experienced, there are three characteristic features:

- at any one time and place, föhn winds are a by-product of the general wind direction. If the wind direction changes by 180° the distribution of weather conditions either side of the mountain range is reversed. For example, if a westerly wind blows across Scotland, a location such as Aberdeen on the east coast may enjoy the föhn effect. If the wind direction changes to an easterly, places on the west coast such as Fort William and Oban will have these conditions (Roy, 1997);

- shallow hills, rising at least 50 m above surrounding land can generate noticeable föhn conditions, although not so dramatic as the effect experienced to the lee of larger mountains (Stone, 1983);

- globally, the frequency of the föhn phenomenon is difficult to establish but it is estimated that it occurs on about 50 days per year; this is based on data from the European Alps. The highest frequency and magnitude of the föhn wind occur in stable conditions when mountains (rather than convection) are responsible for forcing air upwards. Warm poleward-moving air is more likely to be stable and this explains, for example, why the northern rather than the southern side of the Alps is affected.

SUMMARY

The Sun provides the energy that drives the atmosphere and the global climate system. In the lower atmosphere, air temperatures are determined locally by the balance between incoming and outgoing radiation (i.e., net radiation), and by the Earth's surface properties because the atmosphere warms from below.

Air is either in a stable or unstable state depending on its temperature gradient (profile). The upward movement of a parcel of air is associated with a change in temperature as its volume changes in response to the varying mass of air above. If it warms and expands, becoming more buoyant than the surrounding air, an unstable state is established; if it cools and contracts (volumetrically) and become less buoyant (denser) than the surrounding air, an unstable state is established.

Temperature changes affect the air's ability to hold water vapour: as cooler air holds less moisture, and air tends to cool as it ascends, it often becomes saturated and is unable to hold any more vapour. Precipitation takes place if sustained condensation occurs; this can happen in either stable or unstable air. If air descends, it tends to warm and thus its capacity to 'hold' water in a vapour state increases. All this activity is strongly influenced by topography and differential warming of the Earth's surfaces.

Understanding the processes discussed in this chapter begins by observing the sky, and particularly the clouds. But, before examining the 'view from below', the next chapter investigates the transformation of heat into motion and the development of weather systems.

REFERENCES AND GENERAL READING

Barry, R.G. and Chorley, R. 1998: *Atmosphere, weather and climate.* 7th edition. London: Routledge.

Drake, F. 2000: *Global warming – the science of climate change.* London: Arnold.

Fischer, A. and Staehelin, J. 2003: The Antarctic ozone hole: 1996–2002. *World Meteorological Organisation Bulletin* 52, 264–69.

Intergovernmental Panel on Climate Change 2001: *Climate change 2001: the scientific basis. Contribution of working group 1 to the Third Assessment Report of the Intergovernmental Panel on Climate Change.* Houghton, J.T., Ding, Y., Griggs, D.J., Noguer, M., van der Linden, P.J., Dai, X., Maskell, K. and Johnson, C.A. Cambridge: Cambridge University Press, 881 pp.

Kiehl, J.T. and Trenberth, K. 1997: Earth's annual mean global energy budget. *Bulletin of the American Meteorological Society* 78, 197–208.

Quaile, E.L. 2001: Back to basics: föhn and chinook winds. *Weather* 56, 141–45.

Roy, M. 1997: Highland and Island Scotland. In Wheeler, D. and Mayes, J. (eds) *Regional climates of the British Isles.* London: Routledge, 228–53.

Smithson, P.A., Addison, K. and Atkinson, K. 2002: *Fundamentals of the physical environment.* 3rd edition. London: Routledge.

Stone, J. 1983: Circulation type and the spatial distribution of precipitation over central, eastern and southern England. *Weather* 38, 173–77 and 200–205.

Thompson, R. 1998: *Atmospheric processes and systems.* London: Routledge.

2

The dynamic atmosphere – energy, motion and the creation of weather systems

Atmospheric energy provides the key to understanding patterns of weather and climate: we notice this energy when we sense a sea breeze or a gale. The catalyst for air motion and the source of this energy is the temperature variation across the surface of the Earth, boosted by additional energy obtained from such processes as condensation, as noted in Chapter 1.

In this chapter we examine how energy is transformed into kinetic energy (wind) through variations in air pressure creating patterns of global winds. The discussion then focuses on how regional patterns of air motion and temperature lead to the development of weather systems (areas of high and low pressure) and how they are shown on weather maps.

2.1 ENERGY IN THE ATMOSPHERE – THE GLOBAL MOVEMENT OF AIR

On both local and global scales, air has a tendency to move to where air pressure is lower. Air pressure can exert a force on air molecules and this can control air motion. This was illustrated in a vertical sense in Chapter 1 by the tendency of a small air bubble to rise because of contrasting buoyancy. Horizontal changes of air pressure are far more variable than vertical changes and give rise to a wide range of air motions. The initial trigger at the global scale can be identified as temperature variations.

INFLUENCE OF TEMPERATURE VARIATIONS ON AIR PRESSURE AND MOTION

Any gas expands as it is warmed; the atmosphere is warmest at low latitudes, where it spreads to a greater height. The depth of the troposphere varies from about 8 km in the polar regions to 18 or even 20 km at the tropics. How does this vertical 'stretching' affect air pressure? Before any horizontal movement of air occurs, the air pressure recorded at the Earth's surface would be uniform, roughly 1000 mbar. Because the atmosphere is shallower at the poles, air pressure decreases more rapidly with height (Figure 2.1).

Air pressure is a function of the mass of air above a given location, acting under the influence of gravity. Figure 2.1 illustrates how a horizontal gradient of air pressure arises from the varying depth of the troposphere; A and B are at the same height but have a difference in air pressure of 200 mbar. This pressure gradient will be present at all heights apart from at the surface (at this stage).

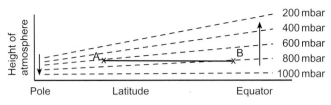

Figure 2.1 *Influence of a pressure gradient in the upper troposphere due to heating at low latitudes.*

FORCES ACTING ON MOVING AIR

The force induced by the horizontal variation in air pressure above the surface is caused by the gradient of air pressure and is therefore called the **pressure gradient force** (Figure 2.2). In reality, as soon as air starts to move it is subject to another force – the **Coriolis force**. The actual path taken by an air molecule as it moves is initially the sum of these forces.

The **Coriolis force** results in a deflection of moving air (when viewed from Earth) to the right of the direction provoked by the pressure gradient force in the northern hemisphere and to the left in the southern hemisphere. This is because we view air motion not from a stationary position but from a rotating surface. We observe the path of air moving as it twists around the Earth's axis of rotation, moving in relation to our position (Box 2.1).

The size of the deflection (the Coriolis parameter) is greater at the poles because the Earth's surface there is at right angles to the axis of rotation. Conversely, it is zero on the equator because horizontal motion there is parallel to the axis of rotation. The size of the force is also proportional to the velocity of the air.

Box 2.1 Demonstrating the Coriolis force

To visualize the Coriolis force we need to recreate the anticlockwise rotation of the northern hemisphere. An easy approximation is to rotate a sheet of paper anticlockwise, marking the centre of rotation as the North Pole. If the paper is steadily rotated with one hand, use your other hand and attempt to draw a straight line towards the pole – in other words, representing air movement acting under the influence of the pressure gradient force. If you only press lightly, allowing the paper to turn, you should find that you have drawn a curve to the right instead of a straight line.

THE OBSERVED DIRECTION OF WIND

The two forces that act upon moving air are summarized in Figure 2.2. As air starts to move in response to the pressure gradient force it is subject to an increasing Coriolis deflection as velocity increases. The result is a westerly airflow (i.e., air moving eastwards). In the southern hemisphere an initial southward (poleward) flow is deflected to the left – this also creates a westerly airflow through that hemisphere. The resulting wind, created at equilibrium between the two forces, is the **geostrophic wind**.

Centrepetal acceleration, a third force that represents changes of airflow direction, acts towards the direction of curvature of the contour lines or isobars shown on weather maps. The curved airflow that is the product of all these forces is the **gradient wind**.

INFLUENCES UPON SURFACE WINDS

Once horizontal motion begins, air is redistributed horizontally and air pressure measured at different locations on the ground will vary. Surface winds are generated in response to this but there are two

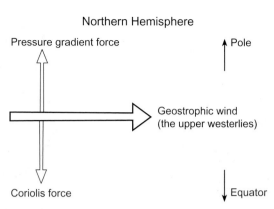

Figure 2.2 *Wind direction in the upper atmosphere of the northern hemisphere.*

important factors to consider: the role of friction on wind direction and the development of surface pressure centres.

Surface winds are slowed down by **friction**. This is especially noticeable over land, which, being rougher than the sea surface, provides more friction. This affects the surface wind direction because one of the contributory forces – the Coriolis effect – is proportional to the velocity of air. The Coriolis force is therefore weakened at the surface by about 30% over land and by about 15% over the open sea. It follows that, while upper level winds blow parallel to the contour lines showing pressure levels, at the surface they blow across isobars of surface air pressure at an angle of 10–20° towards the low pressure centre (illustrated in Figure 2.8).

Patterns of air pressure at the surface are often different to those of the mid- to upper-troposphere. It is simplest to think of separate (but related) patterns of air pressure, one at the surface and the other in the mid- to upper-troposphere. The pressure patterns at these two heights are those most frequently used in weather forecasting. The influence of the troposphere on global winds and weather is now discussed below.

2.2   THE CIRCULATION OF THE ATMOSPHERE – THE UPPER WESTERLY WINDS

FORMATION

We have already seen how the winds above the surface blow from west to east in both hemispheres. This dominant airflow is known as the upper westerlies, which blow from about 500 m upwards, extending from the sub-tropics to within about 15° of the poles. The driving force for this motion is the temperature contrast within each hemisphere. Because of the coldness of the Antarctic land-mass, this difference is much greater in the southern hemisphere (SH), about 60 deg. C compared with about 30 deg. C in the northern hemisphere (NH). The upper westerlies are consequently stronger and more dominant in the southern hemisphere and are impeded by fewer land-masses.

The gradient of temperature across each hemisphere is rarely uniform. This is because of the formation of **air masses**. These are large expanses of air having consistent characteristics of temperature and moisture. Where air masses meet there may be a sudden change (or discontinuity) of these conditions. The rapid temperature change here is an important source of energy for the atmosphere in two related ways:

- it is translated into a local speeding up of airflow, especially towards the top of the troposphere where **jet streams** are identified. These are narrow currents (100–400 km wide) of fast-moving air that encircle both hemispheres. In the troposphere there are two main jet streams in each hemisphere – the polar front jet, which divides polar and tropical air masses, and the sub-tropical westerly jet at about 30°;
- the discontinuity of temperature between different air masses disturbs the relationship between temperature and pressure across each hemisphere. Isotherms (joining places of equal temperature) may not run parallel with isobars (joining places of equal air pressure). Temperature (and hence, the thickness of the troposphere above) will now vary along an isobar. This means that the wind will now blow at an angle to the isotherms, blowing either warm air polewards (warm advection) or cool air equatorwards (cold advection). This is clearly a situation that will throw the temperature pattern into an unsteady state; this is termed a **baroclinic condition**, one that provides energy for the creation of depressions in the mid-latitudes.

WAVES IN THE UPPER WESTERLIES

The upper westerlies in general – and jet streams in particular – have an important influence upon surface weather patterns by determining the direction of airflow at many levels. They rarely encircle each hemisphere in a concentric pattern. The airflow usually shows elements of a waving motion, termed either long waves or **Rossby waves**, after the American meteorologist Carl-Gustav Rossby, who identified their role in the atmosphere in 1939 (Box 2.2).

The larger the waves, the more air migrates between north and south. Figure 2.3(a) shows a situation in which the waves were small and the westerlies kept to a fairly consistent latitude, known

---

Box 2.2 Understanding patterns of airflow; the upper atmosphere, progression and blocking

Airflow in the upper atmosphere is usually shown on maps in terms of the height at which the air pressure reaches a particular value. These are called 'constant pressure' charts; for example, a '500 mbar' chart shows contour lines representing the height at which the air pressure is 500 mbar. The units are decametres; a typical height in the mid-latitudes is around 500 dm, or 5 km above the ground. A low height indicates that the atmospheric pressure is low in that area through the depth of the troposphere and this is usually associated with cold air. The charts are fundamental to understanding the behaviour of the atmosphere because the wind at that height runs parallel to the contour lines.

On average we see a poleward decrease in the height of the 500 mbar level; this reduction is larger in a progressive weather situation (Figure 2.3(a); associated with a large pressure gradient across latitudes) and there is little longitudinal variation in the height. This means that the westerlies follow a direct path from west to east with little meandering through Rossby waves. A more sinuous airflow shape is shown in a blocked airflow type (Figure 2.3(b)) when the 500 mbar height also varies with longitude; poleward extensions of high pressure (greater heights) are **upper ridges** in the Rossby waves and equatorward extensions of low pressure (smaller heights) are **upper troughs**. In Figure 2.3(b) a ridge can be seen over the northeast Atlantic and a trough exists over mainland Europe. The arrows show the generalized direction of motion at around 5 km height.

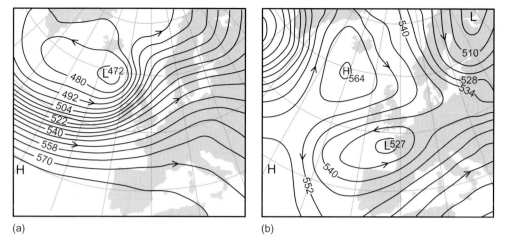

Figure 2.3 *The contrasting upper air circulation patterns on 500 mbar synoptic charts (see text above for explanation of units): (a) a progressive weather situation (10 January 1993); (b) a blocked weather situation (1 January 1997) (the surface weather map and satellite image for this date are shown in Figure 5.8).*

as a **progressive** weather pattern. The opposite case, in which the waves are well developed, is known as a **blocked** weather pattern (Figure 2.3(b)).

The variation between progressive and blocked weather situations is expressed in terms of the **index cycle** (Table 2.1). This varies in a semi-regular pattern, each state possibly lasting for several months. This will affect the distribution of rainfall and temperature over large parts of the hemisphere.

When blocking is well developed, mid-latitude air plunges equatorward in large troughs and poleward in the adjacent ridges. This meridional (north–south) migration of air is an important means of distorting the distribution of temperature (and weather in general) in each hemisphere. This is illustrated by the trough in Figure 2.3(b). In such a trough, cold air from higher latitudes plunges southwards cooling the upper troposphere over Europe. The air here becomes more unstable, encouraging the formation of heavy showers.

Upper troughs have a tendency to propagate cool, unsettled weather. Upper ridges tend to divert the jet stream and thus promote settled (though not necessarily warm) weather. A well developed upper ridge causes the jet stream to inscribe a shape similar to the Greek letter omega (Ω), hence the term **Omega block**. The most extreme situation is for the airflow around large troughs to become cut off from that of the upper westerlies as a whole producing **cut-off lows**; Figure 2.3(b) shows an Omega block over the Northeast Atlantic and a cut-off low over mainland Europe.

Table 2.1 Characteristics of mid-latitude weather in blocked and progressive weather types

Synoptic type	Index cycle stage	Cause	Effects on weather
Progressive	High	Large latitudinal temperature gradients	Increased dominance and strength of westerly winds; stormier in mid-latitudes
Blocked	Low	Smaller latitudinal temperature gradients	Weaker, more variable upper airflow; variable surface wind directions and weather

THE IMPORTANCE OF THE UPPER WESTERLIES

The upper westerlies are important to understanding the weather and climate of much of each hemisphere because they:

- dominate the wind circulation across the majority of each hemisphere, especially in the upper troposphere;
- are an important means of redistributing heat from low to high latitudes;
- influence the route taken by surface high and low pressure systems. This is determined by the size, location and movement of the Rossby waves embedded in the upper westerlies. Regional and local patterns of temperature and rainfall can therefore be related to the upper westerlies.

The link between the upper westerlies and the patterns of airflow and air pressure found at the surface is discussed below.

2.3 SURFACE WINDS AND PRESSURE BANDS

The pattern of winds and pressure that we actually experience at the Earth's surface is modified by the effects of rising and falling currents of air. These are associated with – and interact with – bands of high and low air pressure. These variations of air pressure are superimposed upon the theoretical poleward reduction in air pressure found in the upper troposphere (Section 2.1). The effects of this vertical motion on air pressure and winds in different latitudes can now be examined.

TROPICAL AND SUB-TROPICAL WINDS

The link between the heating of the surface and air motion is most clearly demonstrated within the tropics. The tropics can be defined as those parts of the world that lie within the latitudes that experience an overhead sun for some of the year – i.e., between the Tropics of Capricorn and Cancer.

Uplift of warm air is focused on the meteorological equator – the latitude that has the sun overhead at any particular time of year. This rising air disperses (diverges) to north and south as it encounters the tropopause (Figure 2.4). Divergence in the upper atmosphere is then counterbalanced by convergence of air close to the surface, to replace the air that has risen. Conversely, the upper level air starts to cool as it migrates towards each pole and consequently descends back to the surface at the sub-tropics, around 30°N/S. The result is two cells of circulating air on either side of the meteorological equator. This is the thermally driven **Hadley cell**, named after George Hadley who first identified this feature in 1735.

Equatorial uplift does not occur uniformly but is concentrated in clusters or cells of convective cloud, clearly visible in geostationary satellite images (Figure 2.5).

The focus of convection at the meteorological equator, where surface air converges and starts to rise, is called the **Inter-Tropical Convergence Zone** (ITCZ). Because air is rising and then dispersing horizontally from this latitude, air pressure measured from the surface here will be relatively low, forming the equatorial low pressure zone. The rate of reduction in air pressure with height will be slower than at higher latitudes because of the vertical stretching of the atmosphere (the ITCZ is thus a zone of high pressure in the upper-troposphere).

Away from the ITCZ, under the descending air of the sub-tropics, high pressure is created by the accumulation of air in the upper atmosphere. This gently descending air becomes warmed and dries out as it falls towards the surface, resulting in very low amounts of cloud and the world's maximum receipt of solar radiation (Figure 1.4). It is here that we find the large sub-tropical deserts, such as the Sahara. The juxtaposition of low pressure over the tropics and high pressure over the sub-tropics creates a pattern of prevailing winds – the Northeast Trade Winds of the northern hemisphere and the Southeast Trade Winds of the southern hemisphere.

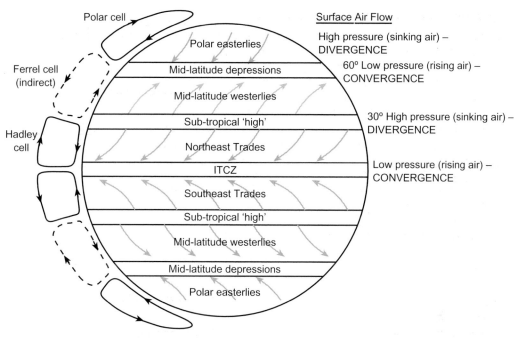

Figure 2.4 *Horizontal and vertical movement of air at different latitudes.*

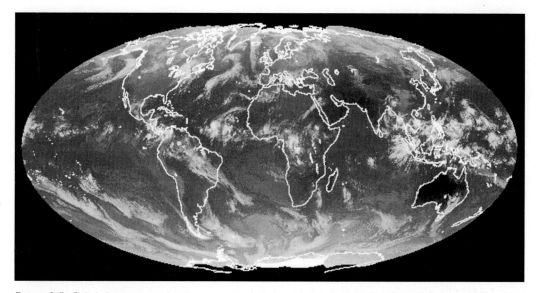

Figure 2.5 *Global climate zones shown on a composite infrared image for 06:00 h (GMT) 15 October 2003. Close to the autumn equinox, the cloud band associated with the Inter-Tropical Convergence Zone (ITCZ) and the mid-latitude depressions of both hemispheres are symmetrically distributed about the equator.*

POLAR WINDS

In the column of air above each of the poles, air is cooling and thus descending towards the surface. This movement starts to create a **polar cell** of air motion in which air at the surface migrates away from the poles, accumulating dense air having created a surface high pressure area. It is later forced to rise in the weather systems of the mid-latitudes, motion which can be visualized as contributing to another complete cell (Figure 2.4). The cell is smaller than the Hadley cell because of the shallower depth of the troposphere at the poles. In this cold environment, it is a low-energy cell and resulting winds are often light and incoherent; the designation of the surface airflow as the polar easterlies is easier to appreciate when viewed in the context of the mid-latitude winds.

MID-LATITUDE WINDS

Mid-latitude winds are extensive in each hemisphere and are an important influence upon weather and climate across a wide range of latitudes. Their movement is determined by two factors:

1 vertically by the descending branch of the Hadley cell over the sub-tropics;
2 horizontally by the behaviour of the upper westerlies, subject also to the influence of surface orography and temperature.

Vertical motion

To complete a series of interconnecting cells, there must be a band of rising air to counterbalance the descending air of the sub-tropics. In reality, mid-latitude airflow is largely controlled by horizontal motions, but it is helpful first to visualize a rising branch of air centred on the zone of temperature contrast in the mid-latitudes that created the polar jet stream. The warmer air mass on the equatorward side of this boundary will rise relative to the colder air mass. This helps create a zone of rising air and the development of a zone of low pressure between about 50 and 60°N/S. This can be thought of as completing an indirect vertical cell, the Ferrel cell (see Figure 2.4).

Although the process is similar to that at the ITCZ, less energy is involved and this energy is obtained from convergence of the air masses rather than convection. It also occurs over a more variable area of each hemisphere because of the important effects of horizontal motion.

Horizontal motion

Horizontal motion in the mid-latitudes is controlled by the upper westerlies. In Box 2.2 the horizontal motion of the upper westerlies was depicted by the index cycle and Rossby waves. The orientation of the polar front jet stream identifies the route to be taken by surface low pressure areas. To identify the actual pattern of surface winds in more detail, we can now examine how a low pressure system is created.

2.4 DEVELOPMENT OF A LOW PRESSURE SYSTEM IN THE MID-LATITUDES

Mid-latitude low pressure systems (depressions) can develop over a large part of the Earth's surface away from the tropics, sub-tropics and polar regions. They are the main source of strong winds and persistent rain in much of the mid-latitudes. They form and develop where (horizontal) temperature contrasts are strong, their path dictated by the upper westerlies for several days, and then they decay as they lose energy.

THE IDENTIFICATION OF LOW PRESSURE SYSTEMS AND FRONTS

The basis of our understanding of these weather systems is the Polar Front theory, developed by a group of Norwegian meteorologists led by Vilhelm Bjerknes in the early decades of the twentieth century (Thorpe, 2002) who created the so-called 'Bergen model' of depression formation. This was the first time that the word 'front' was used, a term originating from the 'fronts' of opposing armies in the First World War. Simple maps of air pressure, temperature and general weather revealed that lines of cloud and rain were often found at the boundary between warm and cold air masses. This led to the definition of a front – the boundary line between air of polar and tropical origin.

Satellite and radar imagery have confirmed the main ideas within this early model, a tribute to the observational skills of the Norwegians who developed it without our present-day technologies for instant global telecommunications and imagery. More modern advances are included in this chapter when they have led to significant changes to our understanding of the processes.

FORMATION OF LOW PRESSURE SYSTEMS

There are two favoured locations for the development of low pressure systems in the northern hemisphere; over the North Pacific around the Aleutian Islands and over the North Atlantic around, or south of, Iceland. Unlike the sub-tropical high pressure systems, these low pressure systems are not permanent features; these areas represent common routes for the track and development of a succession of separate depressions. These areas are characterized by significant temperature gradients (baroclinic zones), not just within the atmosphere, but across the ocean surface as well, where polar and tropical waters meet.

Other parts of the northern hemisphere at a similar latitude do not generate depressions to the same extent because of the influence of land. In winter, the cooling of land encourages the formation of high pressure while in summer low pressure may develop over land in response to surface heating. The situation is different in the southern hemisphere because of the more continuous distribution of ocean leading to the development of a more uniform band of low pressure.

The pattern of the upper westerlies is an important influence on the development of depressions. Figure 2.6 shows how upper ridges and troughs influence the exchange of air between the surface and mid-troposphere. The upper flow is slower at low latitudes because of the weakening of the Coriolis effect. It therefore accelerates as it moves from low to high latitudes between the trough and ridge (northeastwards). This can be visualized as a spreading out or horizontal divergence of the upper-level

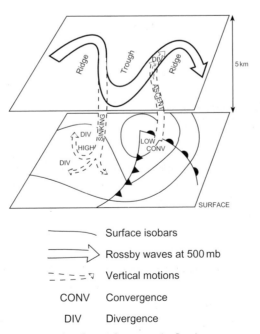

	Surface isobars
	Rossby waves at 500 mb
	Vertical motions
CONV	Convergence
DIV	Divergence

Figure 2.6 *Interaction between upper and surface airflow around a Rossby wave.*

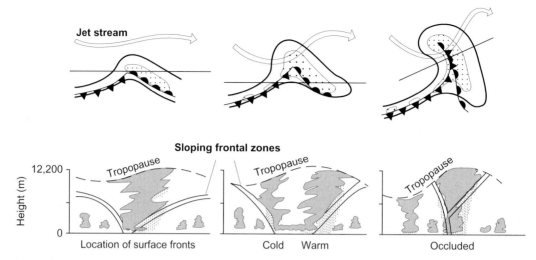

Figure 2.7 *A cross-section through the frontal zones of a developing mid-latitude depression.*

air, lifting up air from directly below to 'fill the gap' (or, more precisely, responding to the reduced air pressure). Upper-level divergence therefore encourages uplift and cyclogenesis. It is counterbalanced by upper convergence (and hence descent) of air flowing from ridges to troughs.

How does air pressure start to fall?

Air pressure is related to changes in the mass of the atmosphere above a location. Changes in air pressure measured at ground level are controlled by the balance between the divergence of air in the upper atmosphere and the convergence of air at lower levels. Figure 2.6 illustrates a situation where divergence is followed by convergence – i.e., the mass of air starts to fall first, hence air pressure falls. Pressure will continue to fall while divergence of air provides the stimulus for air to converge and rise. Air pressure will only rise again if the divergence is less than the convergence of air, though uplift and divergence can also be stimulated by heating of the air.

IDENTIFYING FRONTS IN A DEVELOPING LOW PRESSURE SYSTEM

In vertical cross-section, a front is the surface boundary between cold polar and warmer tropical air masses located where a frontal zone reaches the ground (Figure 2.7).

Convergence forces the warmer air mass to rise as a result of a lower density relative to the colder air. Air that is forced to rise is also forced to cool. Continued uplift may cool the air to below dewpoint temperature resulting in a line of cloud, and often later rain, along the line of the front. Once a low pressure area has started to form, the converging winds do not blow straight towards the front but circulate in a spiral pattern. This is because the wind direction at the surface responds to the new centre of low pressure, around which air will circulate in an anticlockwise direction. The resulting circular motion (as viewed from above) will now start to move different parts of the front and hence the different air masses.

Figure 2.8(a–d) shows the 'life-cycle' of a typical depression. Depressions may lose energy if the original thermal contrast weakens. Diminished energy for uplift will reduce uplift in the system, weakening divergence (output of air) in the upper-troposphere. If convergence at the surface (input of air) now exceeds divergence, air pressure rises and the depression starts to lose identity.

Figure 2.8(a) shows the original front as a simple linear boundary; if the colder air mass is advancing, the boundary is termed a cold front (cold air is replacing warm). If this motion is resisted in one

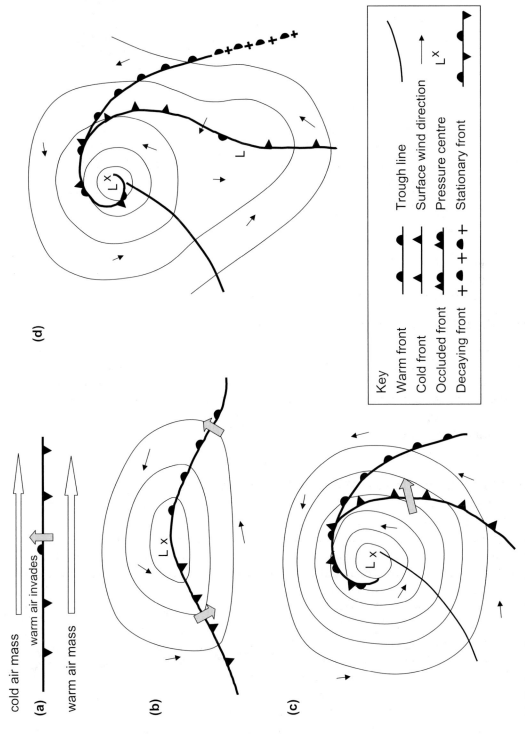

Figure 2.8 *Evolution of a typical mid-latitude depression.*

(a)

cold air mass

warm air invades

warm air mass

(b)

L×

(c)

L×

(d)

L×

L

Key

Warm front — Trough line

Cold front — Surface wind direction

Occluded front — Pressure centre L×

Decaying front — Stationary front

location and the warm air mass advances, the front will move back north, as shown by the single warm front symbol. The term **warm front** is used where the warm air mass is advancing and **cold front** to where the cold air mass is advancing. In Figure 2.8(b) the fronts have become distorted by the anticlockwise rotation of air such that to the west of the new depression centre the front is blown southwards (the cold air advances) and to the east it is blown northwards (the warm air advances). Locations may experience either a drop in temperature at the passage of the cold front or a rise in temperature as a warm front passes. It sometimes happens that winds no longer blow across a cold front of a mature depression (Figure 2.8(d)). As soon as the warm air mass starts to advance, the process of depression formation can start again; creating a secondary depression (Section 2.4).

Frontal zones usually move according to the direction and speed of the winds that blow through them. If the wind stops, the now stationary front may give persistent cloud and rain. If the wind direction reverses, an original warm front will now resemble a cold front (because it is now the cold air that is advancing) and a reversed cold front will resemble a warm front.

The strength of rising air currents within both warm and cold fronts may be denoted by the terms **ana** and **kata**; an ana-warm/cold front is characterized by strong uplift of the warm air mass while a kata-warm/cold front is dominated by the descent of the cold air mass.

The centre of a mature depression is often characterized by spiralling cloud-bands associated with the twisting of the occluded front around the depression centre (Figure 2.9(a) and (b)).

WEATHER SEQUENCES AROUND MID-LATITUDE DEPRESSIONS: UNDERSTANDING SURFACE PRESSURE CHARTS

The aim of this section is to explain how weather changes initiated by a developing depression can be recognized from maps of the air pressure at the surface and from satellite imagery. Weather changes can be illustrated by two examples, highlighting differences between winter and summer situations.

Weather changes associated with the passage of fronts: a winter case study

Figure 2.10(b) shows a surface synoptic chart for 6 January 1998. The warm front is moving northeast across Wales, southern England and France. The cloud ahead of it corresponds with the frontal zone that is slanted forwards. The satellite image (Figure 2.10(a)) locates the forward edge of the cloud from Scotland to Germany. As the warm front approaches, the frontal zone is lower and this leads to sheets of medium-level cloud (such as altocumulus or altostratus – see Chapter 3). At midday, rain and drizzle were being reported from Northern Ireland, southern England and France.

After the warm front has cleared, we are now in the warm air mass in a zone that is aptly called the **warm sector**. This is usually characterized by cloudy, rather humid conditions. There may be drizzle and this can be quite heavy over exposed mountains, whereas sheltered lowland areas may become bright and warm. On the day of the case study, midday temperatures in Cornwall were as high as 12°C compared with just 6°C in London.

A cold front usually has stronger uplift than a warm front and consequently rain tends to be heavier and weather changes more abrupt. The frontal zone is slanted at a steeper angle and the transition from overcast, wet conditions under the front to the arrival of blue sky and bright sunshine can produce dramatic sky changes. The back edge of the cold front cloud often appears as a distinct curve on satellite imagery; this can be seen over southwest Scotland on the satellite image.

The air behind the cold front is unstable because it brings air from a cold source region over a relatively warm surface. Instability at the cold front may be marked by cumulonimbus clouds. Bright sunshine often follows immediately behind the cold front because the uplift on the front is balanced by subsiding air behind it – this accounts for the clear slot in Figure 2.10(a) over Ireland. The instability

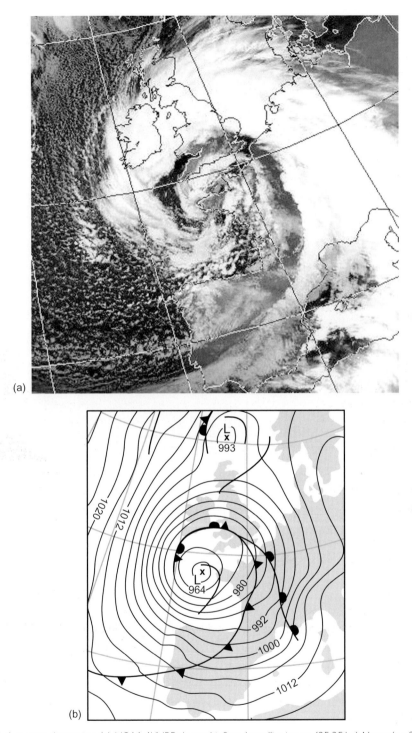

Figure 2.9 *A mature depression: (a) NOAA AVHRR thermal infrared satellite image (05:35 h 6 November 2000); (b) surface synoptic chart for 12:00 h. This is typical of depressions that do not generate extensive conveyor belt cloud, revealing the spiralling vortex of frontal cloud twisting around the depression centre.*

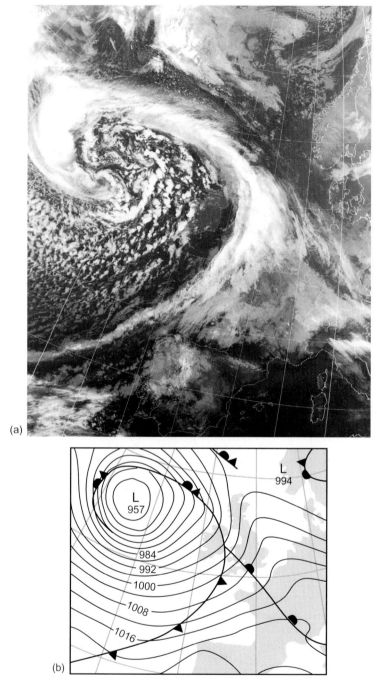

(a)

(b)

Figure 2.10 *A winter depression: 6 January 1998: (a) NOAA AVHRR thermal infrared satellite image for 14:18h; (b) surface synoptic chart for 12:00h.*

soon allows thermals to rise and the resulting blobby cloud pattern of cumulus clouds can be seen to the west of Ireland. The wind direction here is southwesterly but this is air of polar origin, originating from north of Newfoundland. It is this cold origin that sparks the instability as the air runs over warmer surfaces as it is forced southwards around the depression.

In winter, the warmth of the ocean surface moderates the coldness of the polar maritime air. Indeed, the air behind a cold front after crossing an ocean in winter may be warmer than the air over land surfaces. It is still correct to call the front a cold front since the name applies to the temperature change through the depth of the troposphere in general, at heights where temperatures will be unaffected by land/sea temperature contrasts. An additional feature of Figure 2.10(a) is the double structure of the cold front (a 'split front').

A summer case study

Active depressions and cold fronts are less common in summer because of smaller average latitudinal temperature contrasts. However, dramatic cold fronts may sometimes affect western Europe arising from large contrasts in surface air mass temperature. This was illustrated vividly by the ana-cold front of 9 July 1993 (Young, 1994; Figure 2.11(a) and (b)). The warm air ahead of the cold front was subject to strong daytime heating under clear skies. The cold front can be seen as a continuous ribbon of cloud from the depression centre just northeast of Scotland, across Scandinavia and southwest across southern England and Wales. Further north the satellite image shows the characteristic speckled pattern of cumulus cloud. The air is unstable over both land and sea though clearer skies over both the Irish and northern North Sea show evidence for subsidence behind the cold front.

This vigorous front gave up to 40 mm of rain in parts of Wales but was more notable for a rapid drop in temperature. Temperature reductions with the arrival of a cold front can be due to such local factors as the heavy cloud cover, heavy rain and the presence of downdraughts rather than the temperature of the cold air mass behind the front. The temperature in central southern England dropped below 10°C for several hours around midday under the front (the lowest July day temperature since 1920 in Oxfordshire) but recovered by evening as the cloud cleared (Figure 2.12). Temperatures behind the cold front would have remained noticeably colder than those over mainland Europe because of the relatively cool sea surface, unlike the winter case study.

MODERN VARIATIONS ON A CYCLONIC THEME

Recent advances in understanding: explosive cyclogenesis

In recent decades it has been realized that the most active and deep mid-latitude depressions tend to form much faster than typical depressions. Rapidly deepening depressions were found to be associated with a very fast-moving jet stream. The latter is sometimes called a 'jet streak' to denote the fast movement of the air, a symptom of the energy within the atmosphere when large temperature contrasts are experienced across the mid-latitudes. Jet streaks promote rapid uplift of surface air associated with upper level divergence. The resulting explosive cyclogenesis is defined as occurring when depressions deepen at more than 25 mbar in 24 hours (sometimes called 'bomb' deepening).

Seeing conveyor belts of rising and falling air: the contribution of satellite imagery

Understanding the movement of cloud and rain areas within the circulation of a depression remains a major challenge in weather forecasting. Advances in atmospheric observation have centred around explanations of how air rises at a depression and how this uplift is expressed in cloud distributions. During the 1960s and early 1970s observations of rainfall around fronts, using rainfall radar, revealed that much of the precipitation within a depression was associated with condensation within a band of rising warm air in the warm sector – the **warm conveyor belt** (WCB) (Box 2.3). This is an ascending current of air roughly 200 km wide, which flows parallel to, and just ahead of, the cold front (Figure 2.13). It rises from close to the surface (where the air pressure is around 900 mbar) to roughly the middle of the troposphere

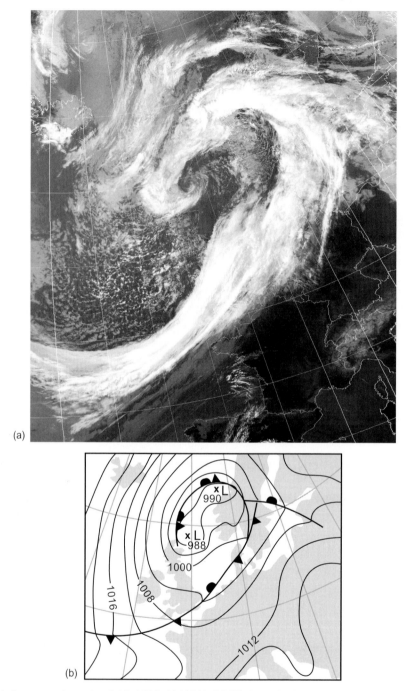

Figure 2.11 *A summer depression: 9 July 1993: (a) NOAA AVHRR thermal infrared satellite image for 08:26 h; (b) surface synoptic chart for 12:00 h.*

at about 500 mbar (or 5–6 km). This is responsible for the deep and extensive cloud seen around major frontal systems and explains why widespread rain may sometimes be observed within the warm sector.

A second band, the **cold conveyor belt** (CCB), feeds colder air towards the depression centre from above the warm sector (Figure 2.13). This airflow produces a distinctive cloud pattern,

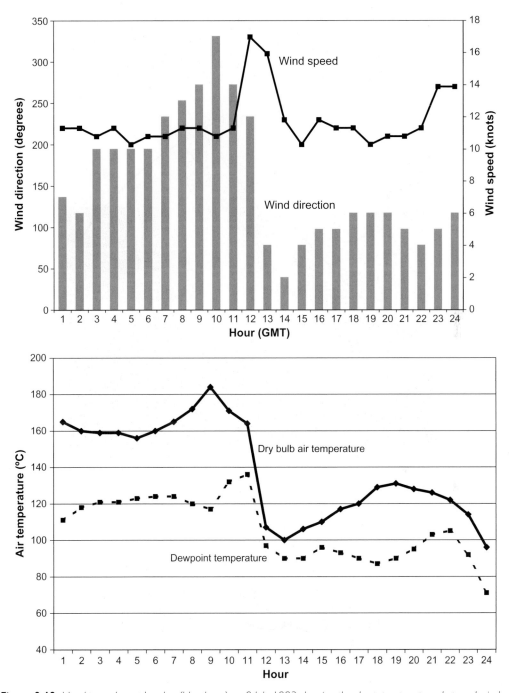

Figure 2.12 *Hourly weather at London (Heathrow) on 9 July 1993 showing the sharp temperature drop and wind changes caused by the passage of the cold front.*

the **cloud head**, that breaks through to the poleward side of the depression (Browning, 1994). The cloud head is also fed by a secondary warm conveyor belt (WCB2 in Figure 2.13). The cold conveyor belt splits into northern and southern branches, the latter inducing the descent of cold, dry air from the upper troposphere and lower stratosphere, creating a sharply defined zone

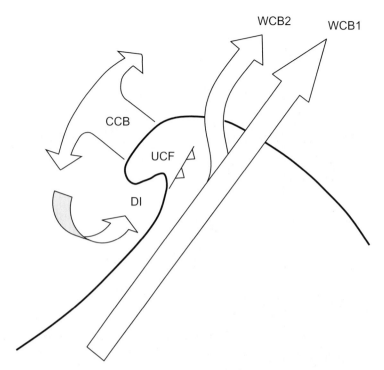

Figure 2.13 *Patterns of warm and cold conveyor belts (of rising and falling air) around a mid-latitude depression.*

Box 2.3 A large warm conveyor belt

A low pressure system developed in February 2003 with a particularly extensive area of cloud associated with the warm conveyor belt. On 8 February this cloud was in the central North Atlantic and the British Isles was under the cloud of a previous depression (Figure 2.14(a)). By late evening the WCB cloud had reached 20°W showing the characteristic fuzzy appearance often shown by rapidly developing cirrus cloud on the forward edge of a warm front.

By evening of the 9th, the WBC cloud had become a large feature just west of the British Isles. Figure 2.14(b) shows strong convection close to the depression centre. The system reached maturity on the 10th with strong convergence uplift over a large area extending beyond northwest Spain. Behind (west of) this cold front a distinct subsidence gap appears as an area of clear skies before the main area of convection cloud appears further away from the front (Figure 2.14(c)). Note the darker tone of this cumulus compared with the frontal cloud, indicating lower (warmer) cloud-tops.

By the 11th, the depression was losing energy. The cold front stalled over eastern England for two days before edging back west. Eastward movement was prevented by the presence of a mass of colder air over continental Europe, an illustration of how a blocked front may 'go into reverse'.

Figure 2.14 *A large warm conveyor belt cloud system, 8–10 February 2003, viewed by Meteosat: thermal infrared images for (a) 10:00h on 8 February; (b) 10:00h on the 9th; and (c) 00:00h on the 11th. (d) A water-vapour image for 06:00h on the 10th. This sequence of images shows how fast cloud progresses in the mid-latitude westerly circulation and the rate at which it develops and decays. The water-vapour image shows how the distribution of water vapour through the atmosphere corresponds with the moist conveyor belt cloud bands (light tone implies moisture).*

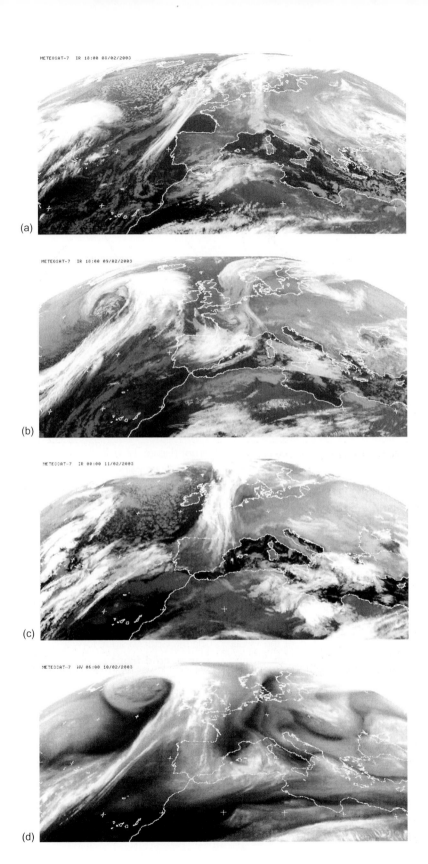

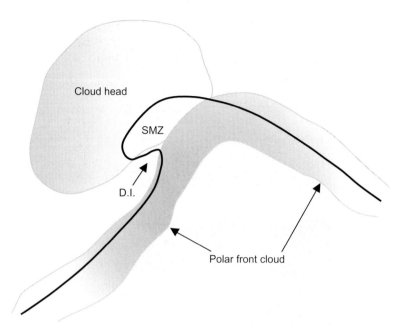

Figure 2.15 *The location of the cloud head, polar front cloud, dry intrusion (D.I.) and shallow moist zone (SMZ) in relation to conveyor belts shown in Figure 2.13.*

of low cloud amount and little moisture – the **dry intrusion** (also called a **dry wedge** or **dry slot**). Notice how this distorts the shape of the cold front (Figure 2.13). It is conspicuous as an area of dry air on water vapour imagery (see Figure 2.14(d)). This air contains gusty winds – having its origins near the polar jet stream – and reinforces the storminess of the developing depression (Browning, 1997). This zone is now associated with the so-called **sting jet** of strong winds in explosive cyclogenesis.

The conveyor belts interact differently according to their relative movement and the strength of uplift (Semple, 2003). If the air in the warm conveyor belt is not progressing as fast as the cold conveyor belt, the WCB air rises strongly above the CCB rearwards, forming an ana-cold front, such as on 9 July 1993 (Figure 2.11). If the WCB remains in contact with the surface, it may be lifted abruptly where it meets the surface cold front, associated with a narrow line of convection (**line convection**) which can appear clearly on rainfall radar imagery as a narrow line of heavy rain.

In other situations the descending CCB air may be progressing faster than the WCB, producing a kata-cold front. This air can push over the WCB, forcing the latter to slope forwards. The leading edge of the cold, descending CCB air is then termed an **upper cold front** (see Figure 2.15); this usually occurs in the surface warm sector, ahead of the surface cold front (this 'split front' is identifiable on satellite imagery (Figure 2.10(a)). This air rides over a **shallow moist zone** near the ground and then rises as it curves round towards the depression centre (Figure 2.15). The presence of dry, cold air over warm, moist air at the surface may induce thunderstorms (Browning and Roberts, 1994).

Cloud signatures of cyclogenesis and explosive cyclogenesis: the contribution of satellite imagery

Analysis of high-resolution satellite imagery in the 1980s and 1990s revealed distinctive features of cloud patterns around a depression (Bader *et al.*, 1995; Barry and Carleton, 2001; Semple, 2003). These cloud 'signatures' are particularly conspicuous when a depression deepens rapidly and can

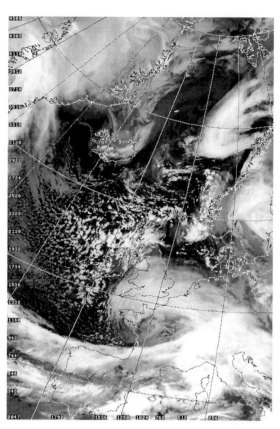

Figure 2.16 *NOAA AVHRR thermal infrared image (19:01 h 27 December 1999) showing a deep depression and a distinct cloud head feature. This system was one of three that caused gale damage in France around Christmas 1999 (see Section 5.2).*

therefore be useful aids to forecasting explosive cyclogenesis. Three cloud patterns are recognized:

1. **Leaf cloud**. A convex swirl of cloud that arises from the anticlockwise rotation of air around a young developing depression; a deformation of a formerly straight frontal cloud. This cloud signifies the condensation of air within the warm conveyor belt.

2. **Comma cloud**. The cyclonic (anticlockwise) air circulation changes the shape of the 'leaf cloud' into the characteristic comma shape of a developing depression. The term is also used to describe the much smaller clusters of convective cloud that may appear on the poleward side of the jet stream within polar maritime air (Williams, 2000). These features sometimes catch up a cold front, leading to a feature known as an 'instant occlusion' when heavy showers merge into a more coherent rain band.

3. **Cloud head**. Cyclogenesis of a mature depression often culminates with this smooth, semi-circular bulge of cloud that forms polewards of a depression centre. It will appear as cold cirrus on a satellite image, being part of the cold conveyor belt (Pearce *et al.*, 2001; Figure 2.16). The feature is associated with vigorous cyclogenesis in depressions forming downwind of an upper trough, the favoured area for depression formation.

Cloud heads, dry intrusions and warm conveyor belt cloud are important signatures of major depressions and are visible on many of the case studies of mid-latitude storms in Chapter 5.

SECONDARY (OR WAVE) DEPRESSIONS

Secondary depressions can be a source of frustration for weather forecasters and a potential source of weather hazards. A secondary depression is a wave-like disturbance on a front, most commonly found on cold fronts (Figure 2.8(d)), occasionally on warm fronts. It represents the formation of a new depression on the front of a mature depression (and is hence secondary to this parent depression).

When the 'host' front is a cold front, it first has to be stopped by the resistance of the rising warm air mass. Enhanced local uplift (perhaps stimulated by local heating) contributes to a local loss of air pressure and inflow of air towards the front. This motion can push the front into reverse, creating a warm front. Continued progress of the cold front on either side of this feature enlarges the resulting wave shape. The wave then 'ripples' along the host front, towards the centre of the mature depression on a cold front and away from it on a warm front. It follows that they form most readily where winds across the front are light: as winds flow in, they start to spin.

Secondary depressions pose an interesting challenge to weather forecasters. The delayed arrival (and clearance) of rain often occurs where the intensity of the rain is enhanced, because of stronger local uplift. The appearance of a convex shape to the back edge of cloud on a cold front on a satellite image can give a clue to the emergence of a secondary depression. The benefits of imagery in near-real time are highlighted by the early identification of such a feature over remote sea areas, long before they reach land. However, successful forecasting of secondary depressions remains a challenge because of their capability for rapid development and because of the local weather contrasts that may develop in their wake (Box 2.4).

2.5 HIGH PRESSURE SYSTEMS – ANTICYCLONES

ORIGINS AND BASIC CHARACTERISTICS

High pressure systems are areas where air accumulates and sinks, leading to settled weather with light winds, a low probability of rain and variable surface temperatures. This air motion balances, and is partly in response to, the rising air currents of low pressure systems. Semi-permanent high pressure systems are found in the sub-tropics of both hemispheres and at the poles (Figure 2.4). They extend their influence into the mid-latitudes where ridges occur in the upper troposphere.

High pressure systems are formed where air converges (horizontally) in the upper troposphere. This is usually driven by the large-scale circulation of air as different latitudes experience either rising, dispersing air (low pressure) or sinking, accumulating air (high pressure). The downward motion can also be driven by the cooling of air over large land-masses in winter, the most notable location being over Asia. In this case the air accumulates in response to this downward motion.

Box 2.4 Regional weather contrasts from a secondary depression

On 16 September 1997 a mature depression was located off the Norwegian coast (Figure 2.17). Where the front across Northern Europe was moving south it was plotted as a cold front (such as over Ireland). In two places it was moving north and was therefore designated as a warm front here. The secondary depression was the larger northwards ripple on the front, crossing Scotland at 12:00h. The location of the secondary served to prolong and intensify the rain over western Scotland and to delay its arrival further south. 24-hour rainfall totals reached 54 mm at Greenock, near Glasgow, a significant contrast to the fine, warm weather over southeast England. More than 10 hours of sunshine was observed across southern England where the temperature reached 22°C.

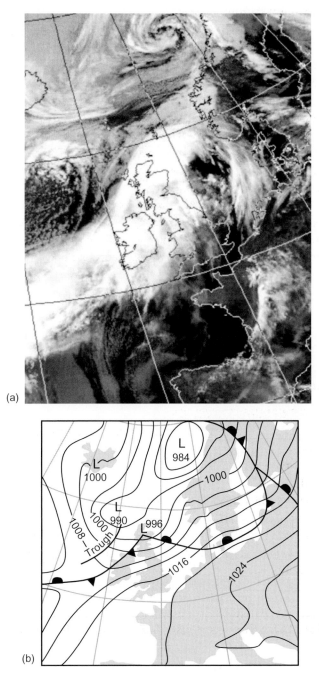

(a)

(b)

Figure 2.17 *Regional weather contrasts at a secondary depression: (a) NOAA AVHRR thermal infrared satellite image for 07:39 h; (b) surface synoptic chart for 16 September 1997. Variations in exposure and shelter had a large effect on the range between the minimum and maximum temperature. This varied from only 1 deg. C on the coast of Lancashire to 10 deg. C in Northumberland and 14 deg. C in southeast England. The reason for this lies in the distribution of cloud and the wind direction. Northwest Britain was cloudy owing to the uplift of air around the secondary depression and because of the exposure to the moist southwesterly winds. This cloud suppressed the temperature range and onshore winds contributed to the nocturnal warmth.*

Air pressure rises as air accumulates, in response to the increasing mass of air. This leads to a second important motion that characterizes high pressure systems and determines their weather: air sinks towards the lower troposphere as it accumulates because large-scale uplift is inhibited by the capping action of the tropopause.

High pressure areas can develop within either cold (polar) or warm (tropical) air masses but the end-results often differ. This can be demonstrated if a cold front clears away a high pressure area in a warm air mass and is replaced by one in a cold air mass. The colder air mass typically has more unstable air, allowing freer dispersal of pollution away from the surface. This leads to improved visibility and a much fresher feel to the weather, partly because of lower dewpoint temperatures (and hence lower relative humidity). If the air within a cold anticyclone becomes more stable, sheets of cloud may become quite extensive and their patchy nature can be difficult to forecast more than a few hours in advance.

SEA-BREEZES AND SEA-BREEZE FRONTS

Sea-breezes are an important aspect of the climatology of anticyclones and can result in distinctive cloud formations that are readily identifiable on satellite imagery (Damato et al., 2003). They are generated by the heating of land surfaces relative to the sea and therefore occur most often in spring and summer. Rising thermals over the land reduce the air pressure slightly (forming a 'heat-low'), allowing moist, cooler air from over the sea to be drawn several kilometres inland, usually from mid-morning to late afternoon. They can only form if the surrounding winds are light (less than about 5 knots). Sea-breeze fronts form at the leading edge of the sea-breeze, a focal point for uplift and hence the development of cumulus cloud (Simpson, 1994). These often stand out clearly on satellite imagery as distinct lines of cumulus in contrast to the clear skies in the stable, cooler coastal air (Figure 2.18(a)).

Sea-breezes can occur in other weather situations when land warms in sunshine and winds are light. There is one situation when they do *not* form in anticyclonic weather: when rising thermals are capped by temperature inversions (discussed below).

TEMPERATURE INVERSIONS AND THEIR INFLUENCE UPON LOCAL AND REGIONAL WEATHER

A temperature inversion occurs where temperature rises with increasing height, 'inverting' the usual temperature profile in the troposphere. This can occur in two locations:

- a **surface inversion** (or radiation inversion) occurs commonly on fine nights when winds are light. As the Earth's surface loses heat by long-wave radiation, the surface becomes colder than the air above;
- a **subsidence inversion** (or anticyclonic inversion) occurs when air descends between two levels above the ground. Adiabatic compression warms the air at the DALR ($0.98°C$ per 100 m) (Figure 2.19). The formation of a high pressure system allows this motion to occur on a large scale, often extending for thousands of kilometres. It is not true to say, however, that high surface temperatures are generally associated with anticyclones. The warming, sinking action does not necessarily extend as far as the surface if the latter air cannot be displaced. In this case, a layer of warm, dry air accumulates at a specific height in the troposphere, the air lower down remaining undisturbed (and thus often cooler). Air at 4 km may sink faster than air at 2 km and air at the surface will not sink or warm at all (though it may be displaced, allowing the warmed air to reach the surface). The result is an increasingly stable temperature profile.

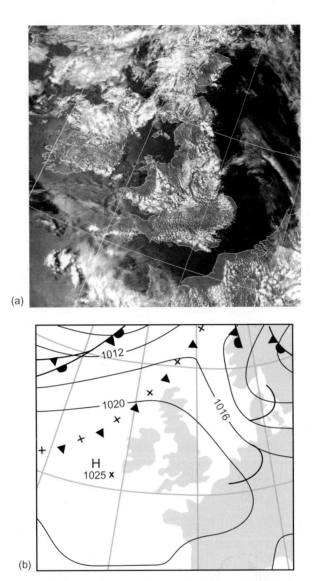

(a)

(b)

Figure 2.18 *Coastal clearance of cumulus cloud caused by the development of sea-breezes: (a) NOAA AVHRR visible image for 15:00 h 19 July 2000; (b) surface synoptic chart for 12:00 h. The contrast between the clear skies over the coasts (in the cool sea-breezes) and the well developed cumulus along sea-breeze fronts is shown along the north coast of France and over parts of England and Wales.*

A subsidence inversion can have the following effects on local and regional weather.

- The presence of warm air above cool makes the air very stable, discouraging vertical movement in the air. This 'cap' on thermals can discourage the formation of sea-breezes when the inversion is at a low level.
- Cloud is dispersed (evaporated) in the warm, dry layer above the inversion layer itself. Since the layer is often sharply defined, cloud conditions may change abruptly. For example, if the inversion height is less than that of mountain peaks, the summits may experience bright, warm sunshine while the lower slopes lie in cloud (Pepin et al., 1999). A strong subsidence inversion

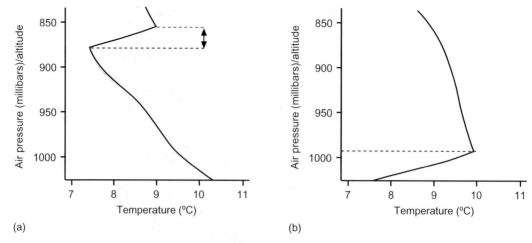

Figure 2.19 *Temperature profile through (a) a subsidence (anticyclonic) inversion and (b) a surface inversion.*

in the winter of 1895 caused the humidity at the summit of Ben Nevis to fall below 10% for several hours in January (Roy, 1997).

- The stable air may be cooled to its dewpoint temperature overnight (or by contact with a cool surface). If sunshine fails to evaporate the resulting cloud, 'anticyclonic gloom' can occur – the persistent cover of low-level sheet cloud (typically stratus) formed below the inversion. Similar conditions develop when radiation fog forms in response to a surface inversion at night. On a larger scale, extensive areas of stratus or stratocumulus may become trapped under an inversion across large cool surfaces where temperature and humidity levels are relatively constant.
- The air below the height of the inversion will be 'sealed off' from the rest of the atmosphere by the stability. Winds are usually light and pollution levels often increase over time. If the air below the inversion is dry, skies may be cloudless despite the intense heating from the sun. Although in this situation air may become thermally unstable immediately above the ground during daytime, the inversion provides a 'cap' to any rising thermals. If this occurs below the potential condensation level in the atmosphere, cumulus cloud will fail to develop (see Figure 1.7).

SUMMARY

Global temperature contrasts, outlined in Chapter 1, provide energy for vertical and horizontal air motion. Air responds to changing temperature by expansion and contraction, which introduces hemispheric pressure gradients. The interaction of a poleward pressure gradient and the Coriolis force create a westerly flow of air through the troposphere in each hemisphere. The strength of this flow varies according to the overall hemispheric temperature gradients; progressive conditions result from higher gradients and blocking tends to develop when gradients are smaller. Blocking allows the westerly flow to meander across a wider range of latitudes. Rossby waves – and their constituent ridges and troughs at high altitude – provide an important impetus to the flow of air between different latitudes, leading to weather anomalies.

The Rossby waves – and the jet streams found within them – dictate the route taken by mid-latitude depressions and provide the essential link with the atmospheric circulation at the surface. Scientific advances, supported by new methods of observation, have identified important links between the weather systems of the surface and the upper troposphere.

The contrasting characteristics of depressions, fronts and anticyclones have been introduced. Further case studies in Chapters 5 and 6 provide more detailed examples of how these characteristics are expressed in different climatic environments.

REFERENCES AND GENERAL READING

Bader, M.J., Forbes, G.S., Grant, J.R., Lilley, R.B.E. and Waters, A.J. 1995: *Images in weather forecasting.* Cambridge: Cambridge University Press.

Barry, R.G. and Carleton, A.M. 2001: *Synoptic and dynamic climatology.* London: Routledge, 620 pp.

Browning, K.A. 1994: Life cycle of a frontal cyclone. *Meteorological Applications* 3, 233–35.

Browning, K.A. 1997: The dry intrusion perspective of extra-tropical cyclone development. *Meteorological Applications* 4, 317–24.

Browning, K.A. and Roberts, N.M. 1994: Use of satellite imagery to diagnose events leading to frontal thunderstorms: part I of a case study. *Meteorological Applications* 1, 303–10.

Damato, F., Planchon, O. and Dubreuil, V. 2003: A remote-sensing study of the inland penetration of sea-breeze fronts from the English Channel. *Weather* 58, 219–26.

Pearce, R., Lloyd, D. and McConnell, D. 2001: The post-Christmas 'French' storms of 1999. *Weather* 56, 81–91.

Pepin, N., Benham, D. and Taylor, K. 1999: Temperature inversions in the Vale of Eden. *Weather* 54, 241–53.

Roy, M. 1997: Highland and Island Scotland. In Wheeler, D.A. and Mayes, J.C. (eds) *Regional climates of the British Isles.* London: Routledge, 228–53.

Semple, A.T. 2003: A review and unification of conceptual models of cyclogenesis. *Meteorological Applications* 10, 39–59.

Simpson, J.E. 1994: *Sea breeze and local winds.* Cambridge: Cambridge University Press.

Thorpe, A.J. 2002: Extratropical cyclones: an historical perspective. In Pearce, R.P. (ed.) *Meteorology at the millennium.* London: Academic Press, 14–22.

Williams, K.D. 2000: Mesoscale analysis of a comma cloud observed during FASTEX. *Meteorological Applications* 7, 129–34.

Young, M.V. 1994: An exceptional summer cold front – 9 July 1993. *Weather* 49, 249–53.

3

Observing weather from the Earth's surface

Observation is fundamental to understanding weather: to 'read the sky' is to understand weather. In the next two chapters we explore how weather is observed, first from the surface and then from space. In the first half of this chapter we will see how instrumental observations of weather variables such as temperature and rainfall are made. The human eye is another valuable instrument for weather observation, particularly changes in cloudscapes. This chapter shows how our view of the sky can be 'decoded' to add to our understanding of the weather.

3.1   THE MAKING OF WEATHER OBSERVATIONS

THE VALUE OF OBSERVATIONS

Observation is the basis of climate science. Observations by instruments and the human eye provide factual data on local weather and it was only through the acquisition of these data from the late eighteenth to twentieth centuries that our understanding of climatic patterns and processes developed. A single climatological station (Figure 3.1) provides a perspective of climate variables over time, the value of these observations increasing the longer the observations continue. It is usually

Figure 3.1 *A typical climatological station.*

thought that records have to be kept for at least 30 years to gain an impression of the climate of that area. The value of the records increases further when we have a network of observing sites in order to obtain a spatial perspective on weather and climate.

INSTRUMENTAL OBSERVATIONS

Instruments used for observing the weather have in many cases changed little since they were invented one or two centuries ago, this continuity being important for the accurate detection of climatic change. After highlighting the use and development of traditional instruments, the automation of weather observation through computer-compatible sensors will be discussed.

Air temperature

Most thermometers work by showing the contraction and expansion of a liquid in a glass tube. A substance that remains in liquid form across a wide range of temperatures and a temperature scale by which we can quantify these changes is required; alcohol and mercury are commonly used. The Celsius scale is calibrated against the freezing and boiling points of water, being at 0°C and 100°C, respectively. The next requirement is that the thermometer records the temperature of the surrounding air rather than the temperature of the instrument itself. Official temperature measurements are therefore made out of direct sunlight.

Following the invention of the thermometer by Galileo in 1597, accurate thermometers first became widespread in the eighteenth century in Europe. As with any innovation, the standard use of the instrument was determined by trial and error. Early thermometers were often exposed in an unheated, north-facing room of a house rather than outside. Despite this, valuable and consistent archives and weather diaries were created at this time (Manley, 1962; Kington, 1997). The problem of providing a shaded but ventilated exposure was not solved until Thomas Stevenson (the Scottish lighthouse engineer and father of Robert Louis Stevenson) devised the Stevenson screen in 1866 (Box 3.1). Variations on this design remain in use throughout the world today (Parker, 1994).

The term 'air temperature' usually applies to the temperature measured at 1.25 m (4 feet), the height of thermometers in a Stevenson screen. The term 'ground frost' refers to frost recorded by a thermometer exposed just above the blades of grass, 1.5 cm above ground level. Because the surface of the Earth radiates long-wave radiation at night, ground frost is more common than air frost that is recorded in a screen at 1.25 m. Soil temperature is measured by thermometers inserted into soil to a variety of depths from 5 cm to 100 cm. As depth increases, temperature variability over time diminishes.

Temperature observation has three attributes:

- officially approved thermometers have an **accuracy** judged against reference instruments to within ±0.3°C across a wide range of temperatures;
- the **resolution** of observations refers to the smallest change in a variable that can be measured or read; it is customary to read to one decimal place on the Celsius scale. High resolution combined with low accuracy can result in spurious accuracy;
- **response time** is the time taken for the displayed temperature to come into equilibrium with the ambient temperature.

Simple temperature measurement can be made using two types of instrument that are cheaper than the official thermometers described in Box 3.1:

- a Six's thermometer (a 'maximum–minimum' thermometer, invented by a gardener, James Six, in 1782) is designed to record the maximum and minimum temperature since the instrument was last read and reset;

Box 3.1 Inside a Stevenson screen

The typical range of instruments housed in a Stevenson screen is shown in Figure 3.2. The two vertical thermometers show the air temperature (the dry-bulb thermometer) and the 'wet-bulb temperature'. The relative humidity is calculated from the extent to which the wet-bulb temperature has been reduced (by evaporation of distilled water supplied from a small container). The greater the evaporation, the lower the humidity. The two horizontal thermometers record the maximum and minimum temperatures since the last observation. A thermograph and hygrograph provide a constant record of temperature and humidity.

Figure 3.2 *The interior of a Stevenson screen: thermograph (on right); thermohygrograph (on left); wet bulb and dry bulb thermometers (vertical, in centre); maximum and minimum thermometers (horizontal, in centre).*

- a whirling psychrometer (whirling hygrometer) is convenient for use in field surveys. The whirling action provides a means of shading and ventilating the thermometer bulbs (the latter increases the rate at which thermometers record ambient temperature after storage).

Precipitation

Rainfall is measured as the depth that would accumulate in a straight-sided cylinder (the depth is independent of the diameter of the cylinder). Precipitation that falls in non-solid form (snow, hail, sleet) is melted before measurement commences.

By the eighteenth century the principles of an accurate rain gauge had been agreed; a circular cylinder with a sharply defined funnel of known diameter and an inner collecting vessel. From the mid-eighteenth to mid-nineteenth centuries, experimental trial and error defined the correct exposure of rain gauges and revealed geographical variations in rainfall.

Standard rain gauges in the UK have a 12.7-cm diameter funnel exposed 30 cm above local ground level and an inner collecting vessel to minimize evaporation. The surrounding area should be short grass. Ideally there should be no obstructions (such as neighbouring buildings or trees) within four times the distance of their height. Various types of recording rain gauges have existed since the nineteenth century and allow the duration of rainfall to be measured. Growth in rainfall observation after the middle of the nineteenth century led to coordination of individual records and many national meteorological services were established at this time (Box 3.2).

Box 3.2 From 'amateur' enthusiasts to the formation of a national meteorological service

During the nineteenth century weather observation matured from being the preserve of individual enthusiasts to a professional pursuit in which the full value of observations was exploited. The invention of the electric telegraph in the 1840s made it possible to transmit weather observations to central collection points. In 1854 the Meteorological Department of the Board of Trade was created, the forerunner of the UK Meteorological Office. A weather forecasting service only developed slowly following a chance disaster. A storm on 25–27 October 1859 sank *HMS Royal Charter* off Anglesey with loss of 500 lives (Lamb, 1991) and this led to the introduction of gale warnings and construction of air pressure charts.

Meanwhile, an assistant in the Meteorological Department, George Symons, inspired by a series of droughts in the 1850s, started to collate rainfall data from observers. He soon realized the potential such data had for enhancing our understanding of the variations in rainfall over space and time. The result was one of the most famous voluntary organizations in meteorology; the British Rainfall Organisation (BRO; Pedgley, 2002). Symons devoted his life to this pursuit and, by the time of his death in 1900, there were 3500 rainfall observers in the British Isles. A century later the number was around 5000. The BRO produced annual volumes of *British rainfall* from 1860. Publication continued after it was incorporated into the Meteorological Office in 1919 but ceased with the 1968 volume.

Sunshine and radiation

The duration of bright sunshine is measured by the Campbell–Stokes sunshine recorder, developed around 1880. The rays of the Sun are focused onto a specially treated card by a solid glass sphere (Figure 3.3). The length of burn indicates the duration of bright sunshine. Errors are likely to occur from the overestimation of sunshine on days of broken cloud cover. Although imperfect, and requiring daily human intervention, continued use of Campbell–Stokes recorders does ensure continuity with previous observations of sunshine.

The measurement of solar radiation has become possible since the twentieth century with the development of instruments such as the pyranometer. The temperature difference between a black and a white surface, caused by contrasting albedo (reflectivity of solar radiation), is measured and this difference can be interpreted as a measure of the strength of solar radiation. Such instruments based on automated sensors can operate as part of computer-driven automatic weather stations and can operate without regular human attention. They can also be calibrated to estimate sunshine duration.

Figure 3.3 *A Campbell–Stokes sunshine recorder. The solid glass sphere focuses sunlight onto a point that produces burn on a card inserted in the frame behind. The length of burn indicates the duration of bright sunshine.*

As a result, the Campbell–Stokes recorder is being replaced at some official observing sites as automation becomes more widespread.

Wind speed and direction

Wind direction is expressed as the direction from which the wind is blowing, usually recorded to the nearest $10°$ from North. In the UK wind speed is officially recorded in knots but scientific research usually measures speed in metres per second ($m s^{-1}$). Official observations should be taken from anemometers on 10 m high masts (e.g., see Figure 3.14) to reduce the frictional effects of passage over surface obstacles. It is usually measured over at least 10 sec in order to record a range of gusts and lulls: it is usual to note both the mean and the maximum gust. The gust:lull ratio may also be quoted, especially in the turbulent wake downwind of obstacles to the airflow (such as a stand of trees or buildings) where the contrast between gusts and lulls typically increases. Similarly, a 'gust factor' expresses the 3 sec maximum gust as a proportion of the mean speed (Linacre, 1992).

TYPES OF WEATHER STATION AND OBSERVATION METHOD

There are two functions of weather stations and two methods by which observations can be made. Sites making observations every hour or three hours provide an input of weather information to weather forecasters, regularly adjusting the inputs to forecast models. These observations also provide a check on the accuracy of weather forecasts and warn of the development of hazardous or unexpected weather. In the UK, stations making observations every hour are called **synoptic** stations – these are often located at airfields. **Auxiliary** stations usually provide observations every three hours.

    Climatological stations provide information about local weather and climate and how it varies over space and time. The climate of each location could be said to be unique to that site and local variations are especially large where abrupt changes in altitude are found. These records assist a wide range of activities such as the planning of agricultural work and the planning of construction projects. Traditionally, observations were made at 09:00 h daily and submitted at the end of each month. This procedure is now changing with the advent of automated observation.

Automated observations

Weather observation moved into a new era at the end of the twentieth century with the widespread use of computer-based automatic weather stations (Strangeways, 2000). This change has been evolutionary rather than revolutionary because of the need to provide a means of observation that is consistent with traditional instruments.

The principle of automated observation involves a variety of sensors connected to a data-logger. The logger is a surrogate for human measurement; the logger collects the data from the sensors at regular intervals and prepares them for computer downloading at a later date. This means that human attention is not required every 24 h, it permits more detailed weather monitoring over time and it allows observations to be obtained from remote sites. Observations are now possible from a wide

Figure 3.4 *A roadside automatic weather station. In addition to measuring air temperature and humidity, wind speed and direction, sensors embedded within the road surface give an indication of road wetness and salt concentration. The observations are remotely interrogated by highway engineers and by weather forecasters.*

variety of sites, from mountain-tops to buoys at sea. One of the commonest applications of this technology is the roadside automatic weather station, installed beside most major roads to give early warning of the hazards of winter weather (Figure 3.4).

Sensors are calibrated to be representative of traditional instruments. Thermometer sensors may be housed in a small screen (a Gill screen) or a Stevenson screen. Most digital or electrical resistance thermometers have a slightly faster response time than mercury thermometers.

One of the largest challenges posed by automated observation comes from the possibility of changing observation time. The convention has been established that daily observations are taken at 09:00 h local time. It is possible to choose any hour when using an automatic weather station and it may be felt that the 'climatological' day should correspond with the calendar day. This would make comparison with traditional stations difficult, threatening the coherence of observation networks.

VISUAL OBSERVATIONS – THE VALUE OF A HUMAN OBSERVER

The most important instrument for the weather observer is the human eye: weather observation is not just about the routine logging of data. A human observer watches the sky, notes how the weather is changing and is alert to unusual phenomena:

- a visual estimate of the amount of cloud cover expressed as the number of eighths (oktas) of sky that are cloud-covered. A note is made as to whether the cloud amount is increasing or decreasing. At a synoptic or auxiliary station, estimates of cloud height and type are logged;
- visibility is estimated according to the maximum distance that can be seen;
- whilst looking at the surrounding area, distant weather types are noted; localized showers or patches of fog may be observed that are not present at the observing site;
- visual observation of precipitation can be valuable if the temperature is between 0°C and +5°C. There is often a mixture of raindrops and snowflakes at these temperatures (sleet). Very small temperature changes can make the difference between rain and snow falling at ground level. Consequently, slight changes of altitude may result in contrasting forms of precipitation;
- the human eye can detect such phenomena as funnel clouds, tornadoes, thunderstorms and optical phenomena such as rainbows and haloes.

3.2   CLOUDS – A GUIDE TO THEIR FORM, DEVELOPMENT AND VARIETY

Clouds are the visible expression of the atmospheric processes of stability, instability and condensation. Watching the variety of cloudscapes will identify whether the air is stable or unstable and help us to understand how the weather is changing.

The variety of cloud types may at first seem baffling. The aim of this section is to show how the naming of clouds with Latin names can reveal their origin and the differences between different cloud types. The language of the sky is partly written in Latin, but it does form a logical storyline if we bear in mind the two key states of stability and instability.

Before discussing the different forms and types of cloud, it may be helpful to consider first the general characteristics of all clouds:

- cloud types are universal yet each cloud is unique. Most types of cloud can be found in all climate zones throughout the world, though the frequency and size of particular cloud types does vary;

- different clouds reflect differing proportions of incoming solar radiation. Since water vapour is a greenhouse gas, clouds also absorb some of the outgoing infrared radiation emitted from the Earth's surface. Hence cloud development usually restricts warming by day and restricts cooling at night, though the net effect on temperature depends on cloud type (on balance, cumulus has a net cooling effect and cirrus has a net warming effect);
- clouds may develop in either stable or unstable air.

CLOUD DEVELOPMENT – COOLING AND CONDENSATION IN AIR

Cloud will develop if air is cooled to its dewpoint temperature. This can occur with or without uplift – and this uplift may either be forced or occur freely through convection. Three methods may be identified:

- cooling by vertical movement (adiabatic cooling) through convection or forced either by uplift over high ground (orographic uplift) or by convergence at fronts;
- cooling by horizontal movement (advection cooling) as a warm air mass may be cooled to dew-point temperature by passage over a cooler surface;
- local contact cooling at night when air is cooled from below through contact with cold land surfaces.

The products of these processes may be either cloud, fog or dew, according to the exact location of the condensation and the method by which it came about; this is summarized in Table 3.1.

The formation and characteristics of the products of condensation can now be examined.

FOG

Advection fog (coastal or sea fog)

Advection fog is formed when air moves horizontally over a cooler surface. It is most common around coasts where there is a large land–sea temperature difference. It also forms extensively over the sea where there is large temperature contrast, notably between the Gulf Stream and the cold Labrador Current off Newfoundland. Widespread coastal fog also forms where moist winds blow across a

Table 3.1 The outcomes of condensation

Type of cooling	Trigger for heat loss	Key processes	Product
Vertical motion	Adiabatic expansion	Convection (unstable air)	Cumulus cloud
		Orographic uplift (in stable or unstable air)	Cloud over high ground
		Frontal uplift (in stable or unstable air) (or convergence uplift)	Bands of frontal cloud
Horizontal motion	Contact cooling	Conduction caused by advection of air over a cold surface	Low cloud or sea fog (advection fog)
Small-scale contact	Local contact cooling	Nocturnal cooling by emission of long-wave radiation	Radiation fog / Dew / Frost on the ground

cold ocean current, for example, the fog that frequently shrouds the coast of California formed by condensation over the cold Californian Current in the Pacific.

In late spring and summer, when the sea is colder than the land, banks of advection fog around the coast will quickly evaporate after passing over the warmer land (Figure 3.5). If the wind is blowing (gently) from sea to land, the coastline may remain swathed in fog all day, resulting in markedly lower daytime temperatures than in the sunnier locations a few kilometres inland.

In polar areas cold air from land surfaces may chill the vapour evaporated from the sea surface to produce **steam fog**. A similar phenomenon is sometimes observed around rivers and lakes after a cold night. Moist air around the water surface is cooled by contact with the colder air above. Another source of local condensation is **saturation fog** which may form in a shallow layer above wet ground following heavy rain.

Radiation fog

On calm, clear nights land will cool because of the emission of long-wave radiation. If the surface cools to dewpoint temperature, condensation may first be seen on the ground – droplets of dew may form on grass and sometimes on tarmac. If the surface temperature is below freezing point, sublimation of water vapour to ice will produce a hoar frost. The air above will then cool by conduction. The longer this cooling continues, the greater the layer of air will be cooled below its dewpoint temperature forming an increasingly deep layer of radiation fog.

Because cold air sinks into valley bottoms and hollows, these become 'reception' areas where air may accumulate if winds remain light. A temperature inversion gradually deepens as the pool of cold air develops. Valleys may fill with radiation fog by dawn but the surrounding hills may be startlingly clear and sunny – lying above the pool of colder, dense saturated air. The air will now be extremely stable with little or no mixing between the warm buoyant air above and the cold air within the fog. The result is a boundary or 'cap' corresponding to a particular height with good visibility above (Figure 3.6).

Satellite imagery is a very effective tool for viewing the changing patterns of fog and for relating them to orography. A good example is that of 23 December 1994 when the hills of southern England rose above a shallow layer of radiation fog (Kidd, 1995; Figure 3.7).

Figure 3.5 *Advection fog crossing Three Cliffs Bay, Gower, South Wales, evaporating as it blows onshore (moving from right to left).*

Figure 3.6 Radiation fog filling the Kennett Valley in central southern England on a winter's morning; the surrounding hills are clear (photograph © J.F.P. Galvin).

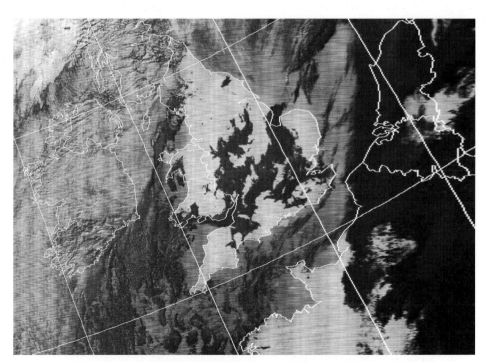

Figure 3.7 NOAA AVHRR (Channel 3) image showing the distribution of radiation fog at 10:57 h on 23 December 1994. The fog is black because it is treated as though it were a thermal image (explained in Box 4.10). The distribution of radiation fog in the valleys of southern England is identified; the inversion was sufficiently shallow to reveal above the fog layer not just the mountains of Wales but also the Mendip, Cotswold and Chiltern Hills.

Dispersal of radiation fog requires a trigger to break down this stable equilibrium. It can be dispersed by an increase in wind speed (leading to mixing of the air) or by solar heating. The effectiveness of the heating depends upon the depth of fog and the extent to which the air has been cooled below dewpoint. Once evaporated, the air rises gently but this may then lead to a temporary re-saturation in the air – the product is low cloud (stratus, discussed on p. 75). This, in turn, may then be evaporated by sunshine. This is why low cloud may temporarily appear after a foggy night before sunshine finally breaks through.

CLOUD FORMATION: THE EFFECTS OF STABLE AND UNSTABLE AIR

The method of uplift and the consequent **cloud form** varies according to whether the air is stable or unstable (Table 3.2).

Figure 3.8 shows the contrast between the cumulus and stratus type clouds that result from unstable and stable air. Owing to the fact that cumulus rarely covers most of the sky, it is usually possible to see the whole of the cloud from base to top. Stratus and other cloud types that occur in sheets reveal only their bases – cloud depth can only be inferred from the darkness of the sky.

The relationship of cumulus and stratus with surface temperature is shown in Figure 3.9. Stratus forms most readily when surface temperatures are low; in winter and at night. By contrast, cumulus develops from surface heating and is therefore most commonly observed at the warmest times of the day.

In the modern era, cloud types (or genera) are classified according to their height; high, medium and low (Figure 3.10). These are then sub-divided into species and varieties (Table 3.3). Galvin (2003) provides a guide to their identification.

CLOUDS OF AN UNSTABLE ATMOSPHERE

When mountains and cliffs in the sky appear,
some sudden showers and storms are near
Cambridgeshire Weather Lore

Unstable air encourages some of the most dramatic and rapidly changing cloudscapes in the atmosphere. Where thermals pass above the condensation level, cumulus cloud can form, providing dramatic visual evidence for the velocity of thermals. If we watch a cumulus cloud for a few minutes we can usually see changes in shape as it crosses the sky; patches may evaporate (they may then appear shredded), other patches develop (turrets at the top of the cloud may be seen to tumble like gently rising fountains – visible evidence for the activity of thermals).

Table 3.2 The effect of stable and unstable air on cloud form

Instability	Method and characteristic of uplift	Resulting cloud type
Stable air	Orographic uplift – Slow Frontal uplift – Slow, widespread	Stratus family and related cloud types – horizontal development dominant
Unstable air	Convection – Rapid, often localized	Cumulus family – vertical development dominant

(a)

(b)

Figure 3.8 *Typical appearance of (a) cumulus cloud of unstable air, and (b) stratocumulus cloud of stable air. The cumulus shows the characteristic flat bases and sprouting profile of the cloud-top in contrast to the flat sheets of cloud typical of stratocumulus.*

As convection is the principal driving force generating these clouds, their development is determined by surface temperature. In general, they form where temperature falls rapidly with height, usually triggered by surface heating, such as on a sunny morning. Cumulus usually starts to form within the first few hundred metres; lower if air is humid, higher if relatively dry.

(a) Stratus

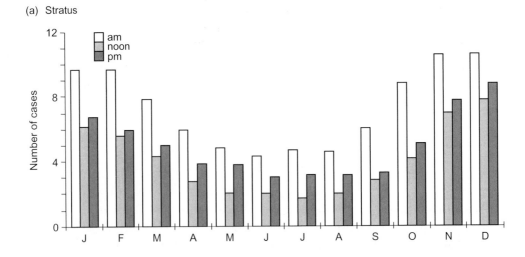

(b) Cumulus

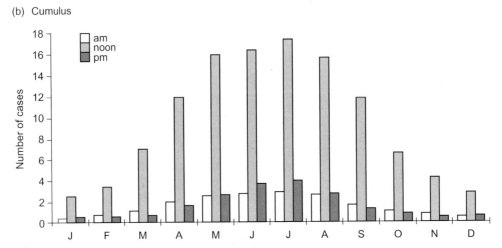

Figure 3.9 *Average monthly frequency of (a) stratus and (b) cumulus cloud at the morning, noon and evening observations at Cracow, Poland (after Matuszko, 2003).*

Cumulus usually first appears as **cumulus fractus** – small, ragged tufts of cloud that resemble shreds of cotton wool. Cumulus forming towards midday will usually have two differences; the cloud-base will often be higher because warmer air is capable of holding more water vapour and will thus have to be raised to a greater height to achieve condensation. The cloud-tops will also usually be higher because the depth of unstable air typically increases towards the early afternoon. *Cumulus fractus* usually changes quickly into one of the deeper forms of cumulus outlined below.

The depth of cumulus is often a good indicator of how unstable the air is. Weak instability, extending over only a shallow depth of the atmosphere, will result in **cumulus humilis** – shallow, thin cumulus clouds of a fine day. If the air is more unstable, small cumulus will readily deepen to **cumulus mediocris**. These may have a triangular shape but the height of the tallest turret of cloud is usually no greater than the width of the base.

We sometimes have several days of fine weather when cumulus clouds resist any response to strong surface heating – they remain as *cumulus humilis*, fair-weather cumulus. They fail to develop further

Box 3.3 The man 'who invented the clouds'

Just over 200 years ago the visual approach to understanding weather made a great contribution to meteorology when Luke Howard, a London chemist, devised cloud names that remain, to this day, the standard international system for cloud identification. Howard devised Latin names for the cloud types that referred to the key attributes and processes of clouds and the surrounding atmosphere. The main cloud forms were identified in 1803 as follows:

Cirrus	streak clouds	high cloud in delicate streaks or lines
Cumulus	heap clouds	a sign of unstable air
Stratus	sheet cloud	a sign of stable air

These basic forms were then divided into species of cloud that, in turn, could be divided into varieties, defining the arrangement or transparency of the cloud (the modern counterparts are listed in Table 3.3). This scheme was launched at a time when London was a major centre of scientific curiosity and Luke Howard soon came to the attention of a wide range of scientists and thinkers (Hamblyn, 2001).

Howard's classification was modified slightly to acknowledge the importance of distinguishing clouds at different heights; in the late-nineteenth century medium-level clouds were named. The special place of cumulonimbus (in having a low base but potentially a high top) was also highlighted; it was listed in an 1887 revision as cloud type nine, hence the expression 'to be on cloud nine'. The advent of photography led to the production of the first *International Cloud Atlas* in 1896.

because of a temperature inversion above the cloud-tops. Inversions may become extensive in high pressure areas where large-scale descent of air offers strong resistance to rising thermals.

Figure 3.11 shows the changes that frequently are seen when cumulus develops on a sunny morning. *Cumulus humilis* may develop into **cumulus congestus**, a thick, tangled mass of cloud of varying tones. These may be several kilometres high but, to be classed as *cumulus congestus*, the height must be at least as great as the width of the cloud.

If the unstable layer is several kilometres deep, rising thermals may be cooled below freezing point. Cloud forming here will consist of a mixture of supercooled water droplets together with ice crystals – the latter give the cloud a fibrous (fuzzy) appearance. The sharply defined turrets of *cumulus congestus* will become smooth and start to spread horizontally creating a **cumulonimbus** cloud (Figure 3.12).

There are two species of cumulonimbus. **Cumulonimbus calvus** is the first stage when the turrets of *cumulus congestus* become slightly smudged in appearance, a clear sign of ice crystal formation. This is an important stage in raindrop formation and showers are now likely (see below). **Cumulonimbus capillatus** is the name given to the more fibrous form of cloud-top associated with widespread ice crystal structures (*capillatus* is Latin for 'hairy'). The frozen (or glaciated) part of the cloud spreads sideways because it has reached the height of the tropopause, 12–18 km above ground. The strong temperature inversion caps the top of the cloud and considerable energy has to be redirected horizontally. The resulting shape is said to resemble an anvil, hence the former description of these as 'anvil clouds' (the shape being termed *incus*).

Understanding Weather

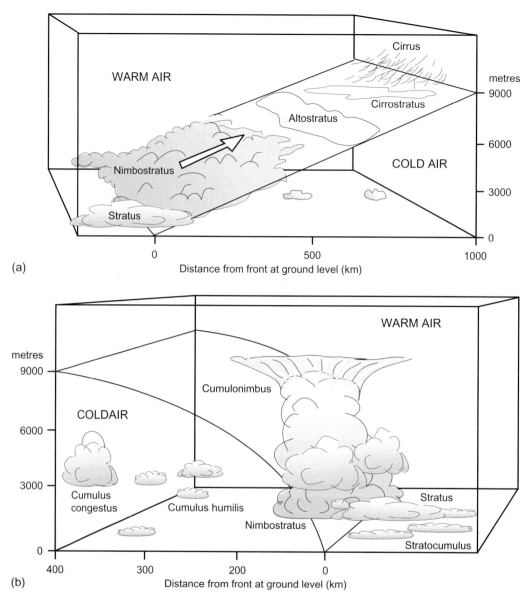

Figure 3.10 *Typical heights of the main cloud types in the mid-latitudes in relation to (a) a warm front and (b) a cold front.*

The distinctive feature of cumulus and cumulonimbus is the speed of development. It is sometimes possible to watch small *cumulus humilis* developing into *cumulonimbus* within a few hours, the shape of cloud changing over seconds. However, the apparent shape of clouds is also a function of our own viewpoint (see Box 3.4).

Showers

Precipitation (rain, sleet, snow, hail) falling from any cloud of the cumulus family is called a shower. Showers usually do not form into large rain areas because of the disconnected nature of cumulus clouds – they rarely cover the whole sky when viewed from below. If the cloud is stationary, a shower

68

Table 3.3 Cloud types and their associated species and varieties. Species describe the shape and structure of the cloud; varieties distinguish between the optical depth (transparency) and arrangement of cloud elements

	Description	Cirrus	Cirrocumulus	Cirrostratus	Altocumulus	Altostratus	Nimbostratus	Stratocumulus	Cumulus	Cumulonimbus	Stratus
Species											
calvus	Smoothening of cloud-top									•	
capillatus	Fibrous cloud-top									•	
castellanus	Turrets at cloud-top	•	•		•			•			
congestus	Towering cloud mass								•		
fibratus	Fibrous form	•		•							
floccus	Tufts	•	•		•						
fractus	Fragments								•		•
humilis	Shallow height								•		
lenticularis	Lens shaped		•		•			•			
mediocris	Moderate depth								•		
nebulosis	Featureless			•							•
spissatus	Dense	•									
stratiformis	Layer		•		•			•			
uncinus	Hooked	•									
Varieties											
duplicatus	>1 layer	•		•	•	•		•			
intortus	Tangled	•									
lacunosus	Regular holes		•		•			•			
opacus	Opaque				•	•		•			•
perlucidus	Gaps in a layer				•			•			
radiatus	Radiating from horizon	•			•	•		•	•		
translucidus	Translucent				•	•		•			•
undulatus	Undulating		•	•	•	•		•			•
vertebratus	Narrow lines of cloud	•									

(a)

(b)

Figure 3.11 *Cumulus development on a sunny morning: (a)* cumulus humilis *approaching the Gower Peninsula, South Wales, from the Bristol Channel, showing the effect of uniform heating across a wide area; (b) much deeper* cumulonimbus capillatus *viewed from the same location one hour later. The growth in the cloud is a function of change in time of day and the warming of land on a sunny morning.*

may be persistent but, in general, showers are short-lived, especially when cloud is crossing the sky quickly. Showers usually form from *cumulonimbus* or *cumulus congestus*. The depth of cloud gives an indication of the intensity of the rain. It is sometimes possible to tell if showers are falling from distant clouds according to whether the cloud base remains clearly visible; if it comes indistinct, it may be obscured by precipitation. Sometimes falling raindrops evaporate before reaching the ground – the 'fallstreaks' visible below clouds are called 'virga'.

(a)

(b)

Figure 3.12 *Cumulonimbus: (a)* cumulonimbus calvus *and* capillatus *showing the slightly softer profile indicative of the start of glaciation but before the anvil shape develops; (b)* cumulonimbus capillatus incus *showing the characteristic 'anvil' shape of the top of the rising glaciated turrets.*

CLOUDS OF A STABLE ATMOSPHERE

In the absence of convection providing energy for rapid uplift, a stable atmosphere presents a scene of more gradual sky changes in which cloud tends to form into sheets or layers which may be either continuous or broken.

Box 3.4 Illusions in the sky?

Perspective plays tricks with our visual perception of cloudscapes. We are used to seeing clouds appear smaller towards the horizon. Imagine the sky is filled with a regular pattern of cumulus clouds (such as in Figure 3.11(a)). Our mind accepts that they might be of similar size even though we actually see them appear to diminish in size towards the horizon. This is one reason why the moon often appears to have a larger diameter when it is a short distance above the horizon.

We can only see the lowest clouds; an entirely different set of clouds may exist above the lowest cloud. The view from above (Chapter 4) may be very different.

Cloud depth introduces a visual illusion when we try to estimate how much of the sky is covered with cloud. If you look at cumulus clouds directly from below or above, they might typically cover about half of the sky. If you look at the same clouds when they are towards the horizon, it will no longer be possible to see any sky between the clouds, because of the height of the clouds. This is why cumulus clouds sometimes appear to gather towards the horizon.

When the sun is overhead on a fine day, the sky appears blue because short-wave visible energy is preferentially scattered by atmospheric aerosols over a relatively shallow (vertical) depth of atmosphere. The sky appears red when the sun is setting and very low on the horizon, because the transmission distance through the atmosphere has increased (approaching the horizontal) and only longer, red wavelengths manage to reach the observer without being scattered. Perceived sky colours are the result of Rayleigh scattering (Box 4.4) with colours becoming increasingly varied when the atmosphere is heavily laden with particulate matter (e.g., from pollution, dust or volcanic activity).

Low cloud – stratus and stratocumulus

Stratus and stratocumulus are found where stable air becomes saturated quite close to the ground surface. They are especially common in maritime, west coast mid-latitude regions such as northwest Europe.

Stratus results from condensation within the first 500 m of a stable atmosphere and is therefore most frequent around windward coasts and hills. It also forms over high ground where air is cooled below its dewpoint temperature by orographic uplift. It often creates overcast conditions making it difficult to see whether there is any cloud above it. In the absence of higher cloud, stratus will give little more than drizzle.

Stratocumulus is another sheet-like cloud that typically occurs between 500 m and 2000 m. As the name suggests, it may have a cumulus type structure. This is misleading and relates only to the appearance of thin stratocumulus rather than its form. Clearly stratiform, on close inspection it often consists of gentle undulations or rolls (*stratocumulus undulatus*). Where it occurs as a thin layer, it may take on a blotchy appearance, revealing small patches of blue sky (*stratocumulus perlucidus*; Figure 3.13) or larger clear areas (*stratocumulus lacunosus*).

Although stratocumulus it is not the most exciting-looking cloud (it often looks like a monotonous grey blanket across the sky – *stratocumulus stratiformis*) it can occur in two distinctive situations that

Figure 3.13 *Well-broken* stratocumulus perlucidus *giving bright weather.*

can have an important influence on local weather. It is often seen towards the end of a fine day when rising cumulus clouds start to flatten out and spread horizontally. This is a sign that the cloud has encountered a more stable layer of air, or that the air is progressively becoming more stable over time. This latter situation can indicate the arrival of a warm front and depression.

Both stratus and stratocumulus may develop by contact cooling over a cold surface – over the land in winter and the ocean in summer.

Medium-level cloud – altiform cloud

There are two medium-level clouds found at about 3–7 km above the ground – altocumulus and altostratus. Again, their names conveniently demote their appearance and height: 'alto' locates them above low clouds whilst the latter names describe the detailed form of the cloud. **Altostratus** is a white or grey veil of cloud typically found where the air is rising because of (frontal) uplift of air and therefore often foretells rain. Altostratus can be sufficiently translucent to show a fuzzy glow where the sun shines through the cloud (*altostratus translucidus*).

Altocumulus is a beautiful cloud that has been aptly described as resembling cumulus ripples where cumulus-like blobs of cloud are compressed into a sheet. It usually has breaks to allow patches of blue sky to be seen (*altocumulus perlucidus*). Thicker altocumulus (*altocumulus stratiformis*) has a regular 'quilted' appearance. The contrast between the white of the cloud elements and the blue of sky can be appreciated to the full when viewed through a polarizing filter of a camera (Figure 3.14).

High-level cloud – cirrus

Above about 7 km, water vapour is relatively scarce but in this cold environment condensation can occur with much less moisture than at higher temperatures. The cloud that forms here is therefore

Figure 3.14 Altocumulus perlucidus *(thin and well broken)*.

thin, wispy and usually composed largely of ice crystals. An important exception is the cirrus-like form of the anvils of cumulonimbus cloud, a product of energetic thermals rising from lower levels of the troposphere in unstable air (dense cirrus, *cirrus spissatus*).

There are three sub-types. **Cirrus** (Figure 3.15) itself typically develops as a thin, lace-like veil of ice crystals. Individual elements may be hook-shaped or irregular, representing the pattern of winds at the top of the troposphere, though linear bands of cirrus are often caused by aircraft condensation trails (Box 3.5). If cirrus becomes more extensive it may be indicative of uplift ahead of a warm front, especially if it is seen to progressively 'invade' the sky from one direction. If this continues, a uniform sheet of **cirrostratus** may form – *cirrostratus nebulosus*. The thinness of the cloud means that it is whiter (optically thinner) than altostratus. When sunlight passing through the ice crystals is refracted, optical phenomena such as haloes and mock-suns may be observed.

Cirrocumulus combines the features of cirrus with the globular pattern of cumulus. The result may be called a 'mackerel sky' but this is also characteristic of *altocumulus*. The elements of cloud with *cirrocumulus* are small and the cloud is well broken. It can be difficult to distinguish between this and *altocumulus* – true cirrocumulus is quite rare (Figure 3.16).

(a)

(b)

Figure 3.15 *Cirrus: (a) at sunset; (b) thick cirrus with patches of* altocumulus, altostratus *and* cumulus *(near the horizon) – looking east from Charmouth, Dorset, towards a retreating depression centre.*

WAVE CLOUDS AND CLOUD STREETS

On occasions an area of cloud may appear to stand still in the sky. On closer inspection, it may appear that the wind is blowing through it, sculpting the cloud into a smooth lens or almond shape. A wider perspective reveals the formation of similar clouds on each side. These are **wave clouds**, forming as stable air passes downwind of high ground (Vosper and Parker, 2002).

(c)

Figure 3.15 *(Continued) (c)* Cirrus intortus – *a tangled mass of rather thick cirrus.*

Box 3.5 Clouds from human activity

The frequency of cirrus in many parts of the world is increasingly because of the presence of condensation trails from aircraft. Aircraft exhaust emissions provide a plentiful supply of condensation nuclei and water vapour. The resulting lines of cirrus can actually give us valuable information about the state of the upper troposphere. The presence of several condensation trails indicates that the air is moist at this level, inhibiting their evaporation. Persistence of contrails can foretell the approach of a front (Kastner *et al.*, 1999). Estimates from NOAA AVHRR imagery give typical coverages of 1% over Europe (Bakan *et al.*, 1994). If they persist they may change their shape and widen (Figure 3.17) according to the upper winds.

Condensation trails may be having an important effect on the global climate system by moderating the daily range in temperature. Like all forms of cirrus, they absorb, scatter and transmit a high proportion of incoming short-wave and outgoing long-wave radiation, reducing the diurnal temperature range (Box 4.8). A three-day suspension of aircraft flights over the USA following the attack on the World Trade Center in New York in September 2001 appears to have coincided with an abrupt short-lived increase in the daily temperature range across the USA. A longer-term trend to lower daily ranges in many areas may be due to increasing air travel.

Ship's trails form in a similar manner to condensation trails but because they are released at the surface they are warmer than those from aircraft. They have a similar appearance on visible wavelength satellite images but they will not show up on thermal infrared images because they have a similar temperature to their surroundings (see Figure 4.8).

Another type of cloud derived from human activity occurs when impressive isolated cumulus clouds are created above thermals rising from power station chimneys – these can occasionally be seen on NOAA AVHRR high resolution images.

Figure 3.16 *A veil of cirrocumulus with a fall-streak hole. Sudden freezing of supercooled water droplets produces ice crystals that occasionally fall to Earth, leaving a clear slot in the cloud sheet.*

The commonest type of wave cloud is composed of perhaps the most beautiful species of altocumulus – *altocumulus lenticularis*. Like all wave clouds, these develop where very stable air blows smoothly over high ground, introducing a wave-like motion over and downwind of the mountains (Figure 3.18). This smooth, lens-like (*lenticular*) cloud forms on the crests of the waves as the air is forced above the condensation level. If this is repeated several times, lines of altocumulus may be seen parallel to the high ground, this overall pattern often being more clearly seen in satellite imagery (Figure 3.19).

Lee waves can form wherever air is stable; if this occurs at higher levels, cirrus may be formed instead of altocumulus. The distinctive feature of lee-wave cloud is that it forms in lines perpendicular to the prevailing wind, i.e., on satellite imagery, if the mountain range initiating the waves is linear in shape, the wave cloud will lie downwind, *parallel* to the mountain range (Figure. 3.19).

If lines of cloud are seen (either from above or below) that are parallel with the wind direction they are probably 'streets' of cumulus cloud (*cumulus radiatus*). These occur when thermals rise in lines. As a line of cumulus develops, the downdraughts descend on either side of the 'street' of cloud, creating linear bands clear of cloud on each side. The illusion of perspective suggests that, when you look *along* the streets, they may appear to converge towards the horizon; when you look *across* the streets, each successive one appears smaller and the clear sky between the cloud lines may be obscured by neighbouring clouds.

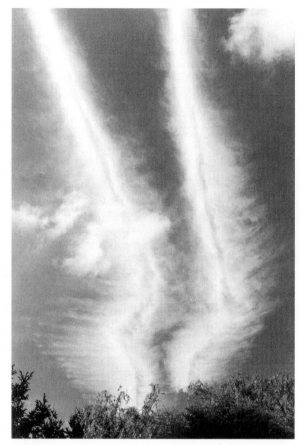

Figure 3.17 *Condensation trails spreading across the sky due to the movement of the upper level winds.*

Figure 3.18 Altocumulus lenticularis – *a nearly stationary wave-cloud viewed from a climatological station in Gower, South Wales. This formed in northeasterly winds as a lee wave developed over the Gower peninsula in South Wales after descending from the Brecon Beacons and the hills of Glamorgan.*

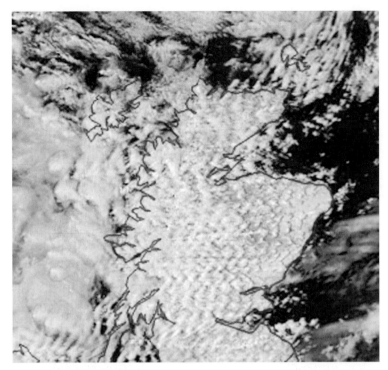

Figure 3.19 *NOAA AVHRR image of lee-wave clouds forming downwind of the mountains of Scotland (16:35 h 18 June 2003). Image supplied courtesy of Mr. Bernard Burton.*

Occasionally, cumulus streets may occur in a shallow unstable layer while lee-wave cloud may be seen in stable layers higher up. Because these features form at right angles to each other, a criss-cross pattern of cloud lines will appear on satellite imagery. This tends to be associated with occasions of vertical wind sheer, when the wind speed in the stable layer is significantly faster than at lower levels. It is even possible for the largest thermals to act like hills by producing barriers to the stable airflow. This generates a wave-like motion many kilometres above the ground, which may then interact with other rising thermals (Bradbury, 1990). Thermals are encouraged when they are beneath a rising wave flow and discouraged by descending flows higher up.

A GUIDE TO CLOUD IDENTIFICATION – A SUMMARY

The myriad of cloud types may appear to challenge our observational skills. The following diagram (Table 3.4) shows how a few simple questions can help identify the cloud types seen on a particular day.

SUMMARY

The dynamic atmosphere provides a constantly changing visual panorama. Observation of cloud from the ground reveals a great deal about how the atmosphere is changing at any one time. We can see immediately whether the air is stable or unstable, whether cloud is developing or dispersing and can gain some idea of how the weather may change. Clouds are the 'engine' of the atmosphere; they are the visible evidence for the transformation of heat energy to atmospheric moisture and motion. Clouds are always changing, responding to the varying stimuli of temperature, wind speed/direction and moisture supply.

In an age of increasing automation and computer processing, weather observation has become more convenient. However, the information gleaned from visual weather observation with the human eye still

Table 3.4 A simple guide to cloud identification

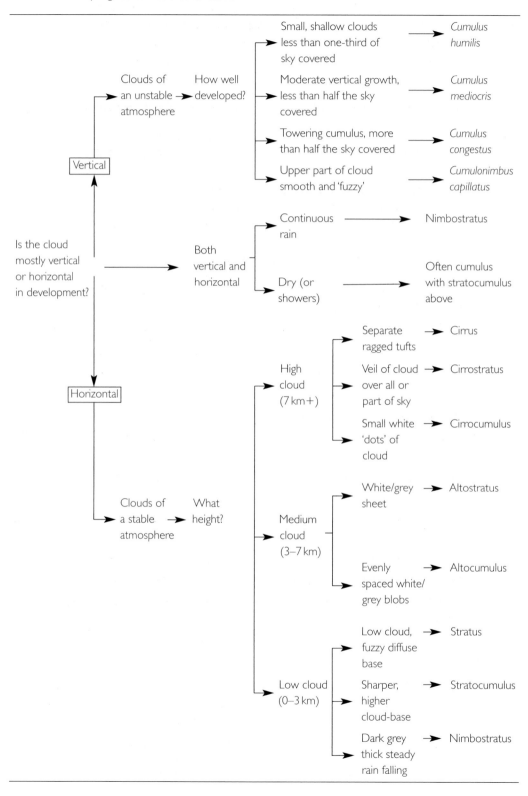

provides a unique insight into meteorology that instruments and computers cannot mimic. It is only by watching the development of a thundercloud that we appreciate how it develops and travels. A sharp eye for falling raindrops can give a useful early warning of the rain turning to sleet and then snow.

In the next chapter, cloudscapes are explored from a more remote perspective in space, well above the weather. This perspective allows us to place the cloudscapes into the context of the development of weather systems outlined in Chapter 2.

KEY SOURCES FOR CLOUD IDENTIFICATION AND INTERPRETATION

Dunlop, S. 2002: *How to identify weather.* London: HarperCollins, 192 pp.
Galvin, J.F.P. 2003: Observing the sky – how do we recognise clouds? *Weather* 58, 55–62.
Ludlum, D.M. 2002: *Weather.* London: HarperCollins.
Meteorological Office 1982: *Cloud types for observers.* London: HMSO.
Scorer, R.S. 1986: *Cloud investigation by satellite.* Chichester: Ellis Horwood.
Scorer, R.S. and Verkaik, A. 1989: *Spacious skies.* Newton Abbot: David & Charles.

REFERENCES AND GENERAL READING

Bakan, S., Betancor, M. Gayler, V. and Grassl, H. 1994: Contrail frequency over Europe from NOAA-satellite images. *Annales Geophysicae* 12, 962–68.

Bradbury, T.A.M. 1990: Links between convection and waves. *Meteorological Magazine* 119, 112–20.

Galvin, J.F.P. 2003: Observing the sky – how do we recognize clouds? *Weather* 58, 55–62.

Hamblyn, R. 2001: *The invention of clouds.* London: Picador.

Kastner, M., Meyer, M. and Wendling, P. 1999: Influence of weather conditions on the distribution of persistent contrails. *Meteorological Applications* 6, 261–71.

Kidd, C. 1995: Images of widespread radiation fog over southern England on 23 December 1994. *Weather* 50, 370–74.

Kington, J.A. 1997: Observing and measuring the weather: a brief history. In Hulme, M. and Barrow, E. (eds) *Climates of the British Isles; past, present and future.* London: Routledge.

Lamb, H.H. 1991: *Historic storms of the North Sea, British Isles and Northwest Europe.* Cambridge: Cambridge University Press.

Linacre, E. 1992: *Climate data and resources.* London: Routledge.

Manley, G. 1962: *Climate and the British scene.* Collins New Naturalist Series. London: Collins.

Matuszko, D. 2003: Cloudiness changes in Cracow in the 20th century. *International Journal of Climatology* 23, 975–84.

Parker, D.E. 1994: Effects of changing exposure of thermometers at land stations. *International Journal of Climatology* 14, 131.

Pedgley, D.E. 1999: Shetland's wake in sea fog. *Weather* 54, 302–10.

Pedgley, D.E. 2002: *A short history of the British Rainfall Organisation.* Occasional papers on meteorological history, no. 5. Reading: Royal Meteorological Society.

Strangeways, I. 2000: *Measuring the natural environment.* Cambridge: Cambridge University Press.

Vosper, S.B. and Parker, D.J. 2002: Some perspectives on wave clouds. *Weather* 57, 3–8.

4

Observation from space: the view from above

Our planet as seen from above, apparently suspended in space and shrouded by huge swirling clouds, is a familiar image revealing dramatically that our world is dominated by weather (Figure 4.1). That we can see this image at all is a product of astonishing scientific achievements. Earth Observation technologies enable us to see familiar sky scenes in entirely new ways, make it possible to visualize the invisible, and help reduce scientific uncertainties that presently reduce the reliability of weather forecasts and climate models. This chapter begins by examining a weather image, moves on to explain the science behind space-based observations, and then considers the view from above.

Figure 4.1 *A cloudy world revealed. The bright swirling, circular cloud mass of Hurricane Isabel can be seen tracking northwest from the central Atlantic towards the eastern seaboard of North America. GOES, thermal infrared, 17 September 2003 (see Isabel at a higher resolution in Figure 5.24).*

4.1   A PERSPECTIVE FROM SPACE

The strong message conveyed in previous chapters is reinforced here: observation is the key to understanding weather. It is about detecting and then understanding patterns and processes, working out the narrative, and evaluating the probability of future events. The sky, aptly described as a battlefield or stage, is the location of conflict and transformation expressed in the dynamic behaviour of clouds (Figure 4.1). From space, it is possible to explore this drama at appropriate spatial and temporal scales, discover critical connections and begin to understand the synergy that is expressed as weather.

Remote sensing and meteorology are equally concerned with the science of **object–energy inter-actions**, albeit using different tools and methodologies. Their strategic coupling has hugely enhanced our understanding of climate and weather.

Weather results from the complex behaviour of energy within an environment loaded with solid, liquid and gaseous matter. Earth Observation devices, remote from this activity, measure the invisible **electromagnetic energy** that escapes to space. This outgoing radiation retains its own history – spectral memories that have to be unravelled before any useful information can be gleaned.

4.2   PATTERNS OF BRIGHTNESS

An image is not the weather: it is a constructed representation of reality containing patterns that connect with atmospheric processes (Figure 4.1). Once deciphered, a huge amount of information is unlocked: a 'picture' is indeed worth far more than ten thousand words, and each one reveals a unique story.

Figure 4.2 captures a weather drama occurring one October morning. In the image we can detect **brightness** (tonal) variations (Box 4.1) because our eyes are stimulated by light energy: recognizing shapes and patterns results from complex neural activity in the brain. Energy, propagated at different **wavelengths** (frequencies), carries 'information' that becomes meaningful once it is decoded. Interpreting a weather image depends upon knowledge of atmospheric processes and the principles of remote sensing: but first the observer must be observant.

IMAGE DECONSTRUCTION

Unravelling brightness-encoded information begins as a visual problem-solving exercise. What is detected (not identified) is a product of our eye–brain system that is amazingly adept at processing 'raw' signals – but only at a very few wavelengths.

First consider what is given – a grey tone image (Figure 4.2) without an informative title or legend. It is essential that the scene is initially 'fixed' in geographical space – here the highlighted coastline is helpful.

Location

The spatial setting of a weather scene is very important and it must always be placed in a global context. Over short time periods, clouds broadly occupy the same location indicating an inherent order in what appears to be a chaotic system (Figure 4.1). Climate scientists seek to explain anomalies, or perturbations, in these 'organized' patterns over longer time periods.

Figure 4.2 *One stormy October day. Shapes, patterns, edges and textures: an image contains a wealth of information but it has first to be decoded. NOAA AVHRR, visible, 30 October 2000.*

Scale

Scale can be gauged once the scene is located. From below, large-scale cloud systems are only partially seen: from space, the entire phenomenon can be observed – but usually just in two dimensions. The central white feature in the image, many times larger in area than the British Isles, translates into a huge amount of cloud energy, most probably associated with dramatic weather. Even in two-dimensional representations, it is easy to appreciate the colossal amount of three-dimensional space taken up by cloud systems and get a sense of the pattern and quantity of stored energy above the Earth's surface.

Brightness and tone

No conscious effort is needed to detect the bright, comma-shaped feature in the scene because we are adept at perceiving strongly contrasting objects, we excel at finding edges and are sensitive to subtle tonal differences. We detect, and distinguish between, weaker and stronger reflected signals (dark and light tones); space-based sensors work in a similar way (Box 4.1). Homogeneous tones (white, black or grey) are associated with relatively uniform spatial conditions while heterogeneous tones (graduated, irregular or mottled) indicate more complex atmospheric conditions.

Box 4.1 Brightness, pixels and grey tones

Space-borne devices are designed to measure 'brightness' or the **intensity of photons** propagated at particular wavelengths. These wavelength 'clusters', described as **spectral bands**, are usually located within **atmospheric windows** (Box 4.5). Signals are **encoded** and transmitted to ground control stations as 'brightness' values using the **binary system**. An **eight-bit system** is often employed constraining brightness variations to 256 (0–255) levels.

Each discrete brightness value represents a defined spatial entity or patch of the underlying scene (see Figure 4.4). Sequences of numerically coded brightness values (digital numbers or DN), represented by **picture elements** (pixels) are arranged to form an image of the whole scene. The encoded values (DN) are converted into a 'pictorial' form using a **grey scale** that, logically, gets lighter as brightness values increase. However, the convention in meteorological **thermal infrared** images is to reverse the logic (darkening as values increase) so clouds remain white, and familiar, even though the brightness signal is very low because cloud-tops are cold (Figure 4.1)!

Digital enhancement involving data manipulation improves the 'readability' of an image but does not change the information locked up in the 'raw' numbers. Colour enhancement, usually employing **false colour** rather than simulated real colours, is entirely cosmetic – the brightness information remains the same. As most people can only distinguish about 20 different grey tones, colour coding is used to 'reveal' more in the image as we can differentiate between many colours and this aids detection and interpretation.

Edges

Edges, essentially virtual edges, are critical to perception and interpretation because they define shapes (entities), textures and patterns, and contain a huge amount of wavelength-dependent (**spectral**) information about what is happening in the atmosphere. An edge is detectable when it is a boundary between adjacent, contrasting tones (e.g., black/white tones produce sharp edges). Sharper edges usually correlate with more extreme contrasts in local atmospheric conditions (e.g., frontal situations; Figure 1.7(a)).

Shapes

Clouds are distinguishable by their shape: shapes gain significance from the space around them. Cloud shapes are time-limited responses to dynamic situations and provide clues to temperature and moisture regimes. If when observing a cloud it becomes less clearly defined as edges become fuzzy, this signals a change in atmospheric conditions.

Detecting shapes, observing the way they change and possibly disappear, is important because it leads to questions about why clouds are certain shapes, why they persist and why they change and vanish at different rates (e.g., speedy precursor clouds heralding a summer thunderstorm). Repeated shapes create textures and patterns that convey information about meso/micro-scale processes. Cloud shapes and patterns are always transient but from space their behaviour is easily monitored. **Time-sequenced images** are routinely animated now and used to create 'virtual' skies.

Patterns

Clouds are observed as patterns, or visual rhythms, at different scales. Image scale, or **spatial resolution** (Box 4.2), determines whether brightness variations are defined as 'patterns' or 'textures'. Patterns are detected as a result of the proximity and similarity of visual elements: size, shape, orientation and

Box 4.2 Resolution issues

In the **Earth Observation sciences** 'resolution' involves space (spatial), time (temporal), brightness (**radiometric**) and wavelengths (spectral). We detect and resolve illuminated objects above a critical size because the eye is sensitive to the intensity and frequency of reflected and transmitted light energy, and excels at locating edges. But the eye's capabilities are constrained spectrally, and by distance and timing factors. Artificial sensors have similar, but less constrained, capabilities. We cannot 'know' what the object is until the spectral information is processed by our brain: identification of an object involves pattern recognition and is an experientially determined construct on the part of the observer in both the real and virtual world.

Spatial resolution

To 'resolve' an object is to distinguish between different brightness levels and separate the object from its surroundings. The probability of detecting an object of a particular size from space depends on the **instantaneous field of view** (IFOV) of the sensing device, i.e., a portion of Earth's surface or atmosphere that is sensed at a moment in time. In an image this is represented as a picture element (pixel) (Box 4.1, Figure 4.3). Whether an object is detected as a distinct entity, i.e., can be spatially resolved, depends on its size compared with the IFOV, its spectral properties and the spectral sensitivity of the detector.

Spectral resolution

When observing a rainbow we resolve, or 'tune in' to, different wavelengths and our brain is able to construct these as separate colours. Objects, like colours, are differentiated by their spectral properties. Sensors are tuned to just a few, or many, **wavelength bands**; each band incorporates a carefully selected wavelength range that may be narrow (high resolution) or broad (low resolution) (Figure 4.4). Sensor stimulation is dependent on the intensity of **incident energy** propagated out to space via atmospheric **transmission windows** (see Box 4.5). Generally, weather images are constructed from broad-band data that are less discriminating because objects of interest (e.g., clouds) are spectrally relatively homogeneous.

Radiometric resolution

Detecting clouds depends on processing brightness information. The degree to which different levels of brightness can be recorded is a measure of the radiometric resolution of a system. Systems are designed to record many brightness levels (low contrast but detailed images) or just a few levels (high contrast images but less detail). Brightness levels are coded as binary digits (bits): the more 'bits' used the higher the resolution or number of brightness levels. Brightness levels are allocated to specific grey tones for image display. The minimum value is 0 and normally represented as black, the maximum value (e.g., 255) as white (Box 4.1).

Temporal resolution

In a moment of sky observation, clouds disappear or shift rapidly out of visual range so the frequency (timing) of observations is critical. Temporal resolution involves the revisit time of a satellite and is determined by its orbital characteristics. Depending on the system used, the same patch of sky can be sensed at high and low resolutions (e.g., every 15/20 minutes, at 12-hour intervals, or every 16 or 26 days). High temporal resolution is essential for weather observations.

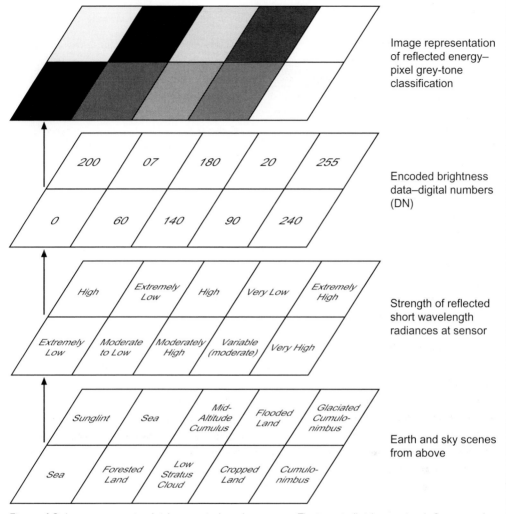

Image representation of reflected energy–pixel grey-tone classification

Encoded brightness data–digital numbers (DN)

Strength of reflected short wavelength radiances at sensor

Earth and sky scenes from above

Figure 4.3 *Image construction: brightness, pixels and grey tones. The intensity/brightness signals from ground cells (that includes the tops of clouds!) are recorded by space-based sensors, encoded as digital numbers (DN) and represented as pixels (the image equivalent of ground cells) in grey tones.*

association bind features together. Here 'pattern' refers to large-scale, repeated shapes associated with distinct cloud entities in **low resolution** images (Box 4.2).

There is a distinct pattern in this scene created by the repetition of a spatially extensive shape, namely, a hooked apostrophe (') or, when inverted, comma (,) feature (comma cloud) (see Section 2.4). In effect, the eye–brain system forms a composite 'picture' by arranging the visual elements – the fragmented edges and tonal contrasts – into a specific configuration or coherent 'comma' shape: this is the essence of pattern detection.

Texture

Clouds are always textured and textures are extremely informative. They convey an impression of motion, or its absence, and contain important clues about processes. In this scene, a variety of

textures can be detected because of contrasting tones (brightness), edges and the close proximity (density) of replicated shapes. Texture perception is a response to the viewing perspective and 'lighting' angle that changes with the Sun's position (a low-angled Sun creates an illusion of greater 'relief'). Some features appear 'smooth' (like silk) rather than 'rough' or highly textured (like orange peel); these are clues to atmospheric dynamics. Knowledge of cloud textures when viewed from below (Chapter 3) can be helpful in interpreting views from above.

Colour

Most images benefit from **colour enhancement**. Simulated 'real' colour (as shown on the front cover) is sometimes employed but false colour is usually used to enhance an image and improve its 'readability' (Box 4.1).

Spatial associations

So far the image has been 'eye-balled', deconstructed and reduced to its major visual elements. A less fragmented approach is also needed.

Creating a coherent 'picture' begins by observing the relationships between visual elements: it is necessary to focus on spatial associations, patterning and the juxtaposition of disparate features. For example, within the main cloud system, different textures and edges are found in quite distinct locations in Figure 4.2. The bubbly texture west of the comma's 'tail' is very different from the feathery texture to the east: this arrangement is an expression of very specific atmospheric processes (i.e., convection and convergence uplift, respectively) and it is all about location.

Cloud systems emerge because energy is transferred in three dimensions. Weather images are usually in two dimensions and this poses some problems. The view from above is of cloud-tops with fragments of the Earth's surface seen through the gaps. It is impossible to accurately determine cloud structure, thickness or altitude because only the top surface layers have been sensed. This has been an impediment to optimal space-based science but new sensors acquire data that enable us to reconstruct clouds in three dimensions (e.g., Figure 6.5) (Rast *et al.*, 1999; Hutchinson, 2002).

Points to mull over

- Mozart was not so much interested in the notes as the space between them: understanding weather is as much about the space between clouds as the clouds themselves!
- Everything visible hides something invisible: seen from above, much may be hidden from the 'eye' in space.
- Cloud scenes are a coalition of visual elements in time and space.
- Weather is the product of synergy: the behaviour of the whole system cannot be predicted from its parts. Components of the Earth–atmosphere system interact to such an extent that changes in the whole affect the nature of the parts, and *vice versa*.
- Vision is an active, creative and participatory process.
- Energy is involved in everything.

4.3   ENERGY: THE INFORMATION SOURCE

There is a lot of it about, it is familiar yet a mystery, and it carries an extraordinary amount of information: it is electromagnetic energy.

Weather is about energy; energy is about **photons**; remote sensing and meteorology are about making connections between phenomena and interpreting the patterns that result.

Images exist because **space-based sensors** detect electromagnetic energy escaping the atmosphere. The story behind this departing energy is far from simple but it is all about photons (imagine these as 'lumps' or packets of energy) and their propagation as waves. Photon 'dramas' occur in the exceedingly thin, heterogeneous gas-rich veil surrounding the planet – but never in isolation from the surfaces below. Making sense of the drama is the key to understanding how we are able to 'see' and learn so much from space.

Weather, and everything else in our world, results from object–energy interactions: solids, liquids and gases can all be considered as 'objects'. From space, visible (e.g., clouds) and invisible (stored heat) phenomena are detected and differentiated because they have distinctive spectral (wavelength) patterns or brightness 'signatures' following object–energy interactions. Accessing the information encoded in signals depends entirely on the human brain deciphering these brightness signatures.

HOT AND COOL BODIES

After about 8 minutes travelling through 150 million km of space, solar energy, essentially a flux of energized particles, drives the coupled Earth–atmosphere system, sustains life on Earth, makes human vision possible and permits the production of images.

All objects, whether solid, liquid or gaseous, emit energy if their temperature is above absolute zero (0 K or −273.15°C). The Sun, the hottest body in our system at around 5800 K, discharges its heat at all frequencies (wavelengths). The output is modelled as the **electromagnetic spectrum** – an energy continuum described in terms of wave behaviour (frequency or wavelength) that is inseparable from photon behaviour (Figure 4.4).

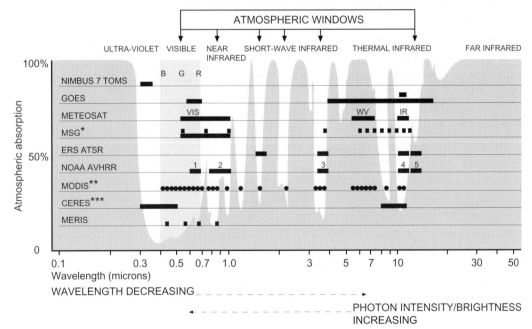

Figure 4.4 *The electromagnetic spectrum. This is a partial view of the spectrum as shorter and longer wavelengths extend well beyond those illustrated. Atmospheric transmission windows (non-shaded) and absorption bands (shaded) are shown, and the link between wavelength and photon intensity (brightness) is indicated. Wavelength (spectral) bands/channels used by selected satellite sensors are located. * MSG – Meteosat Second Generation (2002) has 12 bands; ** MODIS has 36 bands between 0.4 and 14.2 μm; *** CERES also has a very broad band from 0.3 to >50 μm.*

In contrast, the Earth is a cool body emitting less energy at low frequencies (longer wavelengths) predominantly beyond visible light (Figure 4.4; Box 4.3). The atmosphere acts as a buffer zone – an interface between two contrasting energy systems where, as a result of complex energy interactions and feedback processes, an equilibrium state (energy balance) is maintained that finds expression in our weather.

Box 4.3 Emissivity and absorption

All matter, at temperatures above $-273°C$ (0 K), emit electromagnetic energy at all frequencies but the emission pattern (wavelength distribution) depends on the temperature of the object. **Wien's Displacement Law** connects the temperature of an object to the maximum wavelength at which it emits the most radiation. Thus, the cool Earth emits most of its energy at long wavelengths (around 9 μm) and the hot Sun emits most energy at short wavelengths (around 0.5 μm).

Emissivity is a measure of the amount of energy released from an object. It is the ratio of the radiation emitted by a surface to that emitted by a true blackbody (i.e., **Planck's perfect emitter**) at the same temperature (Campbell, 1996: 28). **Blackbody emissivity** is 1. Natural objects, always imperfect emitters or **grey bodies**, have values between 0 and 1. Objects characterized by low emissivity have high **albedos** (e.g., cloud-tops).

Absorbed photons affect the internal organization of an object by increasing molecular or atomic vibrations – an **excitation** process associated with a rise in temperature. Emissions shift to shorter wavelengths as temperatures rise but the rate of change is not linear: a small increase in temperature produces a large increase in heat output. When cooling, emissions progressively shift to longer wavelengths as excited electrons return to their ground state (Figure 4.5).

Cloud-tops at high altitudes are very cold so peak emissions occur at longer infrared wavelengths: lower, warmer cloud emissions peak at slightly shorter infrared wavelengths. Very hot wild fires and active volcanoes emit huge amounts of energy at shorter wavelengths becoming visible and 'coloured'. Ice and snow emit less energy at very long infrared wavelengths. Earth's total energy output is modest; it is a cool body with peak emissions around 9.0 μm. The Sun's output peaks in the visible at 0.5 μm so we see it as a coloured object. Usually, emitted terrestrial energy originating from solar radiation swamps out other sources but the exceptions include geothermal heat from volcanic events, severe fires and nuclear accidents.

Radiative or **brightness temperatures** are not the same as actual temperatures but these can be derived (Gibson, 2000). The actual temperature of water and its brightness temperature are very similar because water behaves almost like a theoretical black body – emitting as much energy as it absorbs. Other surfaces act as grey bodies and their actual temperatures differ from the brightness temperatures.

Emissivity is a complex subject. Factors affecting it include: **spatial geometry** (e.g., Sun angle and curvature of **irradiated** surface); time (e.g., diurnal, seasonal); environmental context (e.g., shadowing); and attributes of the emitting surface (e.g., roughness, colour, moisture content, chemical composition, physical structure, size, shape and orientation). All objects can be characterized by measurable thermal properties, namely, their **heat capacity**, **thermal conductivity**, **thermal inertia** and **thermal diffusivity**. These are measures of an object's ability to store, transmit and release energy relatively quickly (e.g., rocks) or slowly (e.g., water bodies).

The thermal behaviour of objects impacts climate and weather and any human-induced changes in the Earth's system will cause perturbations in its energy budget. The timing and scale of these perturbations are not understood but progress is being made using Earth Observation data in conjunction with ground-based and *in situ* measurements.

Excluding geothermal heat transmitted through the Earth's crust, and local sources such as fires, emissions depend on the absorption of solar radiances. Absorption is a complex process and there is no simple answer as to why different objects absorb different amounts of energy at different rates. All molecules of matter have an initial energy state, caused by internal rotations, vibrations and atomic motion, which is raised to a new, higher level when incident energy is absorbed (Figure 4.5).

The molecular/atomic structure of greenhouse gases (e.g., carbon dioxide, methane) encourages absorption and heat storage. This means that even very small quantities of greenhouse gases have a disproportionate effect on climate and weather. **Water vapour** is particularly significant because it is present in the atmosphere in huge quantities, particularly in clouds.

LUMPS AND WAVES

Earth, a mere rotating speck within the solar system, interrupts some of the Sun's radiation that, once through the quasi-vacuum of space, moves into the particle-rich atmosphere. Incoming photons inevitably encounter objects and collisions continue throughout their temporary sojourn within our system.

Weather can be regarded as the highly organized, rhythmic expression of a complex energy 'dance' that, like all organized systems, adjusts to internal and external forcing factors through the operation of feedback processes. Consequently, the Earth–atmosphere system is not overheating unduly, nor chilling, but sustains a surface temperature of around 15.5°C: this is not a constant and changes may be afoot!

To understand the choreography of the 'dance', it is necessary to understand that photons (think of them as packets or 'lumps' of energy) exhibit both wave-like and particle-like behaviour. It is this so-called duality that is exploited in the design of sensors and it is the key to interpreting images.

Energy has strange properties: paradoxically, it is both contained in very small lumps confined in 'local' space (i.e., photons) and spreads in wave-like motions over vast distances. Imagine photons as 'packets' of pure energy in constant motion. They jiggle around (oscillate) at different rates (frequencies) depending on the level of internal excitation: hotter particles are more 'excited' than calmer, cooler particles. Oscillations vary in response to interaction between electronic and magnetic force fields expressed as wave behaviour – the 'waves' transmit the energy. Our eyes detect moderate frequency waves (**visible wavelengths**) whenever we see objects, including the mechanical oscillations of disturbed water; we detect lower frequency waves when we sense heat (**infrared wavelengths**) emitted from sweaty, jiggling bodies.

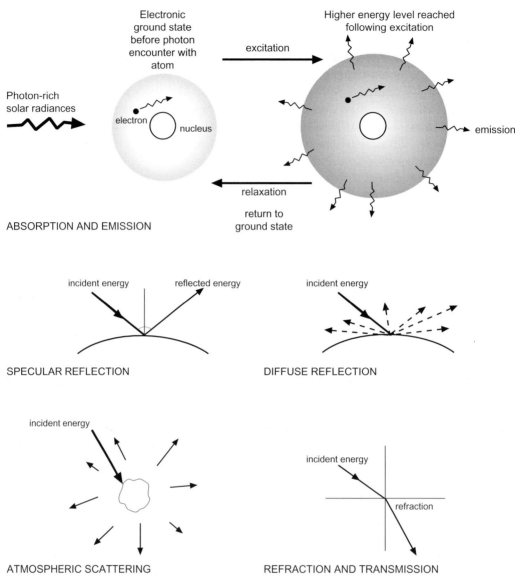

Figure 4.5 *Object–energy interactions: reflection, scattering, absorption, emission and transmission. One process is often dominant: water absorbs most incident energy; cloud-top ice crystals reflect/scatter most incident energy; cirrus clouds reflect and scatter, absorb, emit and transmit energy.*

Solar radiation is not homogeneous. It contains an infinite range of energized particles jiggling vigorously (extremely short gamma waves), moderately (visible waves) and very gently (extremely long radio waves). This **photon-rich flux** experiences little or no interference travelling through space but, as it streams into the denser gaseous atmosphere, **attenuation** (blockage) occurs – mostly because of ozone.

From a meteorological perspective, any long-term reduction in ozone would profoundly affect the climate as more highly energized particles (wavelengths <0.4 μm) would enter the atmosphere: from a biological perspective, this could be catastrophic (Box 1.2). Consequently, the ozone layer is of intense scientific interest and is monitored regularly from space. Less energized photons (>0.4 μm) are transmitted through the ozone layer but interference continues. Most energy reaches the Earth's surface

Box 4.4 Reflection and scattering

Photon paths are perturbed following contact with matter. An object's **reflectivity** is a measure of the ratio between the incident energy and amount of energy redirected away from its surface. Reflection is mostly diffuse but specular reflection occurs under certain circumstances. Natural surfaces generally experience diffuse reflection because they are rough and photons are scattered away from the surface (Figure 4.5); smooth surfaces experience a process akin to 'true' reflection, i.e., specular reflection where the angle of incidence equals, or approaches, the angle of reflection (e.g., **sunglint** off water). Space-based sensors will only detect photons that happen to be redirected towards passing satellites; most will miss.

Scattering is responsible for retaining about 20% of radiances in the atmosphere and 'losing' about 7% to space – especially from high altitude cloud-tops; it occurs because **aerosols** are present in the atmosphere. The relationship between aerosol (particle) size and wavelength is critical in the case of 'selective' scattering. **Rayleigh scattering** occurs when particle size is smaller than the wavelength of incident energy (e.g., gas molecules). **Mie scattering** occurs if particles and wavelengths are about the same size (e.g., smoke particles, very fine dust). **Non-selective scattering** is associated with large particles (e.g., raindrops, cloud droplets, ice particles) and all wavelengths are affected equally. As atmospheric aerosols are predominantly small, shorter wavelengths are scattered vigorously but this decreases rapidly as wavelength increases. The amount of scattering is inversely proportional to the fourth power of the wavelength; consequently, blue light (0.4 µm) is scattered almost ten times more than near infrared (>0.7 µm) but red light (0.65 µm) is only scattered twice as much (Gibson, 2000: 22).

Scattering is responsible for many **coloured sky phenomena** including blue skies and red sunsets; amazingly dramatic, sometimes bizarrely coloured, sunsets follow volcanic eruptions when the sky is loaded with particles.

Albedo

This is a measure of the total amount of energy reflected from an object at all wavelengths. A value of 1 (100%) indicates that all incident energy is reflected and none is absorbed or transmitted; natural surfaces have values below 1. As water absorbs energy, the albedo value can be very low (0.05–0.08) but when transformed to snow it rises to over 0.70. Thick cumulus clouds have average values around 0.75, low stratocumulus around 0.60 and cirrus generally between 0.40 and 0.50. Currently, the albedo value of the **total Earth system** is between 0.31 and 0.33.

Reflections from land surfaces have generally increased following thousands of years of land-cover changes and loss, or modification, of vegetation. The planet must have experienced major albedo changes throughout geological time (e.g., during glacial periods) but human-induced changes, occurring over very short timescales, are being detected that have introduced a new dimension to the energy equation, and our understanding of climate perturbations.

indirectly because photons move every-which-way because of **scattering**, **reflection** and **re-emission** following **absorption**: the same processes occur as energy leaves the atmosphere (Figure 1.2).

Space-based sensors are designed to respond to stimulation by photons leaving the atmosphere and their strength (intensity) is recorded (Box 4.1). Different object–energy interactions produce

---

Box 4.5 Atmospheric windows

Solar radiation interacts with solid, liquid and gaseous matter in the Earth–atmosphere system: this activity is always wavelength-dependent. However, under certain conditions photons are propagated through an apparently transparent atmosphere. Photon behaviour is so 'choreographed' that patterns can be observed and order is recognized in what appears to be a chaotic system. The general pattern is described in terms of atmospheric windows (regions of transmission) and **absorption bands** (regions of attenuation). The spectral bands employed in sensors are determined by the wavelength 'location' of atmospheric windows; absorption bands are avoided except for water vapour detection (Figure 4.4).

Objects are visibly detected when short-wave energy is transmitted through atmospheric windows during sunlit hours; thermal detection depends on the transmission of longer wave (infrared) energy independent of direct solar radiation. Transmission is prevented when photons experience absorption; this depends on a photon's energy level (frequency/wavelength) and the properties of the matter it collides with. Water vapour and other greenhouse gases are highly receptive to less energetic photons (i.e., terrestrial radiances) trapping heat and inhibiting transmission.

Microwave and **radar devices** are used to 'see' through clouds because very long-wave energy is transmitted with minimum interference: this window stretches across many wavelengths and continues into the region of radio waves. All satellite communications depend on extremely long-wave transmissions as communication failures (as a result of interference) occur only under exceptional conditions.

Holes in the ozone layer are 'windows' and entry zones for ultra-violet energy. If not absorbed by atmospheric water, these energy-rich photons represent a serious risk to humans and all living organisms, particularly in low humidity locations at high (and intermediate) latitudes and altitudes.

---

distinctive, identifiable, spectral or wavelength-specific signal patterns or **signatures** that reveal much about what is happening below.

OBJECT–ENERGY INTERACTIONS

Complex interactions occur whenever photons collide with matter. What actually happens depends on the wavelength of incident energy and the properties of matter. The processes involved are described below (and see Figure 4.5, Boxes 4.3 and 4.4).

Absorption

Liquids and solids absorb photons propagated at a wide range of wavelengths: gases are highly selective absorbers. Following absorption, internal agitation occurs and energized electrons 'jump' into new energy levels: higher energy photons cause greater excitation leading to objects becoming hotter.

Absorption bands are wavelength-defined regions where transmission is inhibited by **extinction** due to strong absorption by specific gases (Figure 4.4). Scientists can infer subtle changes (e.g., concentrations of greenhouse gases) by analysing the spectral patterns of outgoing radiances.

Emission

Energy is emitted at wavelengths determined by an object's temperature. Heat is released as electrons relax to their original energy level (ground states) following excitation. Emissions from very hot objects peak at short wavelengths: this explains the colour of heat. Emissions from cooler objects peak at longer wavelengths and colour is absent.

Solids and liquids emit energy at many wavelengths; gases emit selectively. Heat (brightness) patterns can be 'seen' in images because emissions can be detected from space. Actual temperatures of land, sea surfaces and cloud-tops are derived from brightness measurements.

Reflection

Shorter wavelength incident energy generally experiences **diffuse reflection** (multidirectional); under certain conditions **specular** or unidirectional reflection occurs. The albedo of a surface is a measure of average reflectivity that takes into account all wavelengths (Box 4.4).

Scattering

Scattering redistributes photons in all directions but becomes less and less vigorous as wavelengths increase. Basically, scattering is less likely if atmospheric particles are small and wavelengths are long. **Rainfall radar systems** work because artificially produced energy, propagated at very long wavelengths, is **back-scattered** from very large droplets in precipitating clouds.

Transmission

Space observations, and human vision, are only possible because energy can behave as if moving through a transparent medium. Photon paths are perturbed (bent or **refracted**) through optically different atmospheric layers but escape attenuation. Observations from above entirely depend on energy escaping the atmosphere through exit 'zones' described as atmospheric transmission windows where interference is minimal (Box 4.5).

Postscript: uncertain behaviour

Photon behaviour is by no means fully understood. Object–energy interactions are unique, complex and responsive to subtle changes in atmospheric constituents, including those of human origin. Heated scientific arguments, about whether clouds inhibit (by cloud-top reflection) or enhance (by cloud-bottom heat absorption) global warming, indicate the lack of consensus! Space observations, which routinely monitor outgoing radiances, are helping scientists to reduce uncertainty about such issues.

4.4 THE VISIBLE MADE MORE VISIBLE: CLOUDS AND AEROSOLS

The following sections deal separately with weather images constructed from reflected and emitted infrared radiation. In the 'visible' section, clouds and aerosols are discussed; under 'invisible' phenomena (Section 4.5), water vapour and heat energy are examined.

All atmospheric phenomena visible to humans are triggered by object–energy interactions determined by the properties of matter and the wavelengths (0.38–0.72 μm) of incident and transmitted energy. From below, we detect constant shifts in brightness and colours, distinct shapes, patterns and textures because our visual system responds to reflected and scattered photons; but it is always a partial view. What is observed is strongly influenced, and limited, by the eye's spectral sensitivity, the time frame of our observations and our **field of view**: space-based sensors are not so constrained.

Visible images are constructed from brightness measurements recorded during sunlit hours because sensors are stimulated by outgoing reflected energy (Figure 4.6). **Meteosat's VIS** channel incorporates

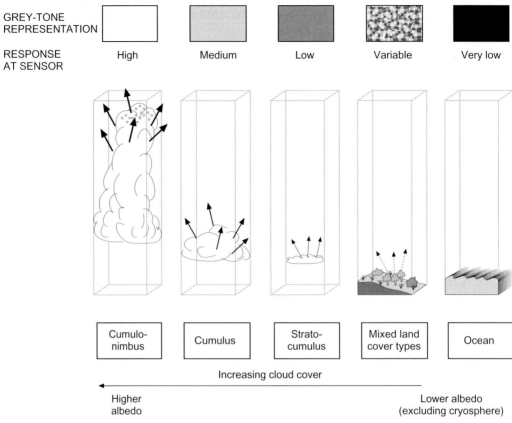

GREY-TONE REPRESENTATION					
RESPONSE AT SENSOR	High	Medium	Low	Variable	Very low
	Cumulo-nimbus	Cumulus	Strato-cumulus	Mixed land cover types	Ocean

Increasing cloud cover

Higher albedo

Lower albedo (excluding cryosphere)

Figure 4.6 *Scenes from above: representation of reflected energy in visible images. A portion of the energy reflected from objects in the atmosphere and from the Earth's surface is detected by space-based sensors. The intensity of the radiation recorded is encoded and represented in image pixels as black (e.g., water), white (e.g., high cloud-tops) and grey tones (e.g., low clouds, land cover).*

visible and the slightly longer, reflected near-infrared wavelengths (0.75–1.1 μm) that are invisible to humans. The latest Meteosat (Meteosat Second Generation launched 28 August 2002) has higher spectral resolution and narrow, more discriminating visible and near infrared channels (Figure 4.4).

On average, clouds occupy over 50% of the global sky and are the 'frontline' reflectors. Most of the redirected energy is no longer available to fuel the Earth–atmosphere systems, but does enable us to observe clouds from above. Aerosols are widely distributed in the atmosphere and scatter energy in all directions. Huge quantities of incoming energy will be attenuated if cloud cover and/or dust-loading of the atmosphere increase dramatically; global images would reveal a complete 'white-out'.

CLOUDS IN VISIBLE IMAGES

Images are constructed from brightness data; brightness is a function of photon intensity as recorded by a detector at particular wavelengths (spectral bands or channels). Broadly speaking, a photon, propagated as a short wave, must have 'made a hit' (encountered matter) and, remaining unaltered, be redirected out to space where it might trigger a response in a passing satellite sensor (Figure 4.6).

Cloud-tops appear bright because they are loaded with water droplets and, at high altitudes, with highly reflective ice particles; atmospheric scattering is reduced in the thinning, less dense air above the clouds. **Cloud aerosols** crucially influence object–energy interactions although their exact role is uncertain (King et al., 1999).

Cloud detection depends on brightness and edges (brightness contrasts). Brightness is a response to a cloud's radiative properties and to external factors. The behaviour of incident photons is affected by a cloud's altitude, **optical thickness**, aerosol properties, water droplet size, ice crystal size and shape, water vapour content and internal dynamics. These properties are highly interrelated and are associated with the cloud's genesis and form (type) and local atmospheric conditions. External factors include the Sun's illumination geometry, platform/sensor specifications and performance characteristics, atmospheric conditions and the Earth's surface radiative properties.

Interpreting visible images: key points (see Figure 4.6)

- Higher, optically thick clouds (e.g., cumulonimbus) are brighter (whiter) than lower, thinner, layer (stratus) clouds (grey).
- Clouds are brighter if water droplets and ice crystals are small.
- Newly glaciated cloud-tops are very bright becoming less bright as ice crystals enlarge.
- Higher, colder ($<-40°C$) cloud-tops (glaciated) scatter more vigorously than lower (warmer) clouds.
- Thin clouds above lower, thicker clouds, dust, snow or ice may be abnormally bright because of spectral 'pollution'.
- Specular reflection from water (i.e., sunglint) may completely swamp local cloud scenes.
- Perceived brightness is enhanced by contrast and this depends on a cloud's optical thickness, what is happening below and whether cloud-free gaps exist:

  - clouds appear 'whiter' above non-reflective surfaces (e.g., dark oceans);
  - relatively dull over moderately reflecting surfaces (e.g., grey vegetated land);
  - 'invisible' above highly reflective surfaces (e.g., white snow).

- Cloud type analysis is based on size, shape, brightness, edge characteristics, texture, pattern, estimated height and spatial context.
- Images constructed from different spectral (wavelength) bands (e.g., infrared and water vapour) need to be explored and compared to discover 'invisible' features.
- Image enhancement can improve contrast and edges; colour coding of brightness values aids detection.
- Subtle changes in brightness/boundary conditions can be detected by inspecting and analysing the numerical (source) data (Figure 4.3).
- Assessing cloud height is problematic as images are two-dimensional representations of three-dimensional scenes:

  - brightness variations may help but the highest clouds are not always the brightest;
  - shadows may give clues to the relative positions of clouds (excluding midday images).

- Wind speed and direction can be estimated from time-series.

The radiative properties of clouds are incredibly complicated, susceptible to rapid change and poorly understood (Box 4.6). However, clouds must be accurately described before their exact role in the energy budget can be determined.

AEROSOLS

In cloud-free scenes aerosols may be visible (e.g., wind-blown dust) or, if invisible, may be responsible for pollution-induced clouds (e.g., **ship trails**). Aerosols are liquid and solid particles suspended in the

Box 4.6 Clouds: the weakest link

A great deal of uncertainty exists about the role of clouds in the climate system because they are extremely complex objects: they represent the weakest link in our understanding (Horvath and Davies, 2001). From space, it is possible to obtain information about cloud location and movements (tracking), cloud height, depth and coverage, cloud layering, composition and internal processes. This information contributes to the scientific understanding of complex feedback processes and the cumulative effect of small radiative changes on global weather patterns. New research findings are triggering new questions – questions that will be addressed through the use of data from recent Earth observation missions employing dedicated systems that have improved coverage of remote oceanic and polar regions.

- *Cloud cover*: changes in cloud cover, detected from space, are forcing scientists to modify their radiation models. Cyclic shifts in cloud cover hugely influence the Earth's energy balance through increased reflection (**negative forcing**) or increased absorption (**positive forcing**). Recent findings suggest that more heat, but less reflected energy, has been escaping the atmosphere from the tropics, rendering current models suspect (Chen *et al.*, 2002; Wielicki *et al.*, 2002).
- *Cloud height*: traditional methods of estimating cloud height using **geostationary** satellite data are inadequate for modelling a complex system. The problem is now being addressed by the latest generation of systems that directly measure cloud height and motion by sensing reflected energy at many different viewing angles.
- *Cloud location*: for the first time in meteorological research, clouds can be located and tracked with precision, improving the reliability of weather forecasts (Horvath and Davies, 2001). Locational information is a critical prerequisite for forecasting severe weather events: knowing the source location of an embryonic storm system, and having the ability to monitor its trajectory accurately, is as important as knowing about cloud height and depth. More accurate estimates of high altitude wind speeds and directions are derived from this data.
- *Cloud type and optical depth*: cloud type determines optical depth. Optically thin clouds (e.g., cirrus) transmit but also reflect and absorb some energy; optically thick clouds (e.g., deep cumulus) strongly reflect and absorb energy and inhibit transmission.

atmosphere (Box 4.7); coarser particles are quickly deposited unless energy levels are very high. A huge number of particles exist in every cubic centimetre of air and are responsible for scattering and absorbing energy. Current research is highly focused on **nucleation** events and the role of aerosols in perturbing the Earth's energy budget, but many uncertainties remain. Human activities are responsible for increasing the particulate loading of the atmosphere and contributing to a whole range of economic, environmental and health-related problems (see Boxes 6.1 and 6.2).

Polluted clouds

At high altitudes, polluted clouds are brighter than non-polluted clouds (King *et al.*, 1999). Aerosols act as cloud condensation nuclei but droplets tend to be small, have longer residence times and scatter vigorously. Globally, precipitation regimes could be affected by increased pollution (especially **atmospheric sulphides**) because droplets do not coalesce and remain suspended in the atmosphere; incoming energy would also be disrupted.

Box 4.7 Aerosols

These are intrinsically linked to the Earth's energy system – and the weather. At a micro-scale, aerosols are components of a very complex system: at a macro-scale it is a hugely complex system. These fine, suspended particles absorb and scatter solar radiances extremely efficiently but no scientific consensus exists about their exact role. Space-based observations, integrated with *in situ* and ground measurements, are helping to reduce uncertainties by providing timely data about **aerosol sources** and **sinks**, spatial distribution, residence time and radiative properties.

Most aerosols originate from natural sources but research suggests that those of human origin are responsible for significant changes in cloud characteristics and the Earth's radiation balance.

The atmosphere is loaded with a heterogeneous collection of liquid and solid particles (0.002 μm–1 mm) of inorganic and organic origin. Coarser primary aerosols include soil dust, sea salt crystals, ice particles, ash, coarse volcanic dust, water droplets and bacteria; finer secondary aerosols include smoke particles, soot, sulphates, fine volcanic dust and smaller water droplets. It is estimated that up to 1 million tonnes of solid particulate matter are suspended daily in the lower atmosphere, mostly at middle latitudes (Kemp, 1994). Dust-loading of the atmosphere has increased as a result of land cover changes, desertification, vehicle emissions and war: an increasingly significant amount of dust comes from burning vegetation (see Box 6.1).

Aerosols absorb, scatter and transmit incident energy and act as **cloud condensation nuclei**: this broad-brush statement disguises the exceedingly complex interactions that occur as photons encounter an increasingly turbid (dusty) atmosphere.

Aerosol optical depth is derived from satellite data. This is a measure of visibility, or how much light passes through a column of atmosphere. Good visibility is typical of clear oceanic skies. Particulate loading of the atmosphere, following industrial, urban and traffic emissions (throughout the world), biomass burning (most continents), volcanic events (geologically determined), dust storms (dry, degraded lands everywhere), sand storms (hot deserts), reduces visibility to extremely low levels and is often extremely hazardous (see Box 6.2).

Dust plumes

Dense, wind-entrained particle fluxes originating from deserts (Figure 4.7), drought-ridden or desertified regions, experience complex object–energy interactions depending on their size, shape and chemistry.

Dust plume detection depends on brightness and contrast, size, shape and texture. Brightness is strongly influenced by optical depth and the radiative properties of the underlying surface(s). Size and shape will be influenced by source/origin and ambient atmospheric conditions, particularly wind speed and direction; textural characteristics are also influenced by wind conditions and local turbulence.

A moderately bright feature in a cloudless scene may be dust. If the dust flux is a broad plume shape in an appropriate geographical context (e.g., deserts) it is reasonable to assume wind transportation of dry particles (Figure 4.7). Narrow, elongated plumes are more likely to be associated with industrial point sources or volcanic explosions. Diffuse, amorphous fluxes (see Figure 6.9) with little contextual information are difficult to identify. The reflectivity of the underlying surface is critical: sunglint may

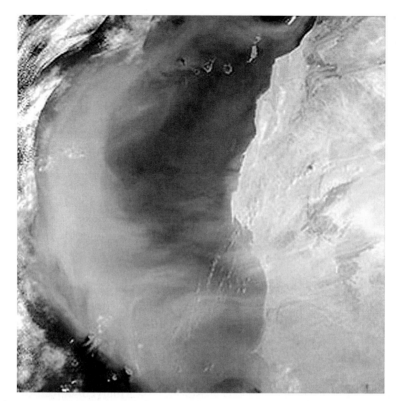

Figure 4.7 *Saharan dust storm. A plume drifts over 1600 km westwards off the coast of West Africa, and then is driven north by the wind, shrouding the sea and the Canary Islands in a thick haze (NOAA/Terra/MODIS, 2 March 2003).*

completely swamp signals from optically thin plumes. If plumes coexist with thin low clouds, differentiation is difficult; widely distributed suspended fine particles produce an amorphous **haze** reducing contrast in visible images.

Interpretation is assisted by general geographical knowledge and by examining animated time-sequenced images (movement and shape changes are distinctive), synoptic charts and thermal infrared images.

Ship tracks

Aerosols (e.g., hydrocarbons) originating from ship exhausts promote condensation (nucleation) and the formation of wispy clouds containing small droplets that promote vigorous scattering.

Tracks appear as distinctive linear patterns often displaying very irregular edges as a result of wind disturbance (Figure 4.8). Tracks may appear brighter amongst low, duller clouds as droplets are smaller. Confusion may arise if sunglint is present or if they coexist with aircraft contrails that are similar, but narrower. Ship trails may be distinguished from **contrails** by comparing visible and infrared images (see below).

Aircraft exhausts (contrails)

Hydrocarbons from aircraft exhausts pollute the atmosphere and may trigger condensation (Box 4.8). Streaky wisps of cirrus-like clouds containing small ice particles develop that vigorously scatter incident short-wave energy (Figure 4.9). These 'contrails' may be difficult to detect if they coexist with 'natural'

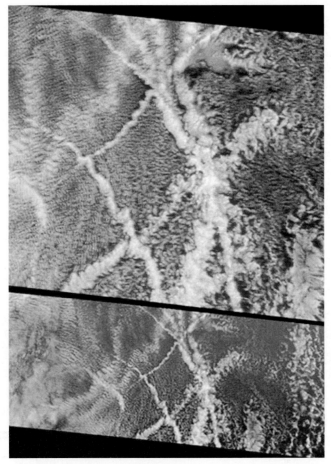

Figure 4.8 *Pollution-induced clouds: ship tracks. The distinctive criss-crossing patterns of puffy clouds can easily be detected embedded in a layer of stratocumulus.* **Stereo anaglyph** *view above, visible image below (Terra/MISR, 18 July 2001. See NASA Website for viewing in three dimensions).*

high, thin clouds because of brightness similarities; distinctive (linear) shadow patterns may aid detection. **Contrail patterns** can be very distinctive in thermal infrared images (see below).

Smog, smoke and soot

Often these aerosols are fairly localized and may be identified by their geographical context (e.g., Los Angeles smog, Arabian Gulf oil fires and smoke from burning forests in Indonesia and Brazil). The frequency and magnitude of **biomass burning** throughout the world, exaggerated, or triggered, in some cases, by anomalous weather, unexpected drought and delayed rainfall, has made this source of pollution an extremely serious problem (see Box 6.1).

Postscript

Images representing reflected energy are visually very informative and excellent for qualitative analysis; behind the image is the **numerical (DN) data** that, once suitably interrogated, are even more revealing. Nevertheless, arguments about the role of clouds and aerosols in the atmosphere are wide-ranging, controversial and contested: all depend on an appreciation of feedback processes operating

Box 4.8 Contrails

During the day, these cirrus-like, artificially induced clouds (Figure 4.9) cover an estimated 0.1% of the Earth's surface. Scientists have known for some time that contrails perturb upwelling and downwelling radiances but there is no consensus about their exact role: the question remains: do contrails contribute to global warming or cooling? The latest sensors deliver more reliable data about the radiative properties of contrails and may provide the answer. Research findings are ambivalent but there are strong indications that contrails contribute to atmospheric warming (Minnis *et al.*, 1999; DeGrand *et al.*, 2000; Meyer *et al.*, 2002).

Contrail incidence is especially widespread over the USA but for three days in 2001 they disappeared following the grounding of most air traffic after the 11 September terrorist attacks. Temperatures in the USA were affected – the range between daily minimum and maximum temperatures increased by 1.1 deg. C on each of the three days (Stenger, 2002; Travis *et al.*, 2002). It is difficult to assess the significance of this 'local' data but they do highlight a need for more precise quantitative research and modelling.

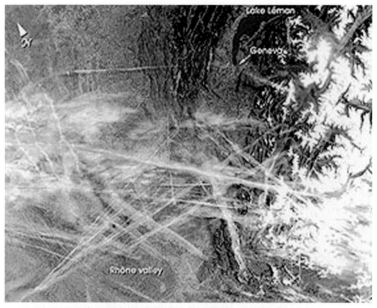

Figure 4.9 *Pollution-induced clouds: contrails streak over Europe (Rhone Valley, France). This is a scene typical of many parts of the world where, under certain conditions, exhausts from air traffic pollute the atmosphere creating these artificial cloud streaks (courtesy of NASA, International Space Station, 15 May 2002).*

within a complex system. New missions, and data from the latest sensors, in conjunction with data from many other sources, are helping scientists to address crucial questions and reduce uncertainties.

4.5   THE INVISIBLE MADE VISIBLE: WATER VAPOUR AND HEAT

From space we can explore the invisible; through imagery we can make the invisible visible. Gases are the invisible ingredients of the atmosphere. Water vapour is arguably the key 'climate' gas and, in company with other greenhouse gases, is inseparable from heat. The invisible is made visible by measuring longer wave, emitted infrared energy. Data collection can be at night because it is not directly dependent on solar radiation.

SEEING THE VAPOUR

Global water vapour (WV) images demonstrate superbly that the atmosphere is a fluid medium in constant motion (Figure 4.10). Hugely energetic air masses, responsible for fuelling the planet's weather systems, meander and swirl gracefully above the planet's surface, their patterns and rhythms revealing much of scientific interest, including:

- tropospheric concentrations of WV;
- structural/dynamical properties (e.g., rotational and deformation features) of cloud systems;
- upper troposphere motions;
- global patterns of stability and instability.

This is all essential information for effective weather forecasting and climate modelling (Browning and Roberts, 1994).

Water vapour strongly absorbs incident radiation at a variety of wavelengths (Figure 4.4) and consequently re-emits heat energy. Spectral data are not obtained through an atmospheric window but via a strong **absorption band** centred around 6.7 μm: WV content and temperatures can be derived from measured emissions (Gibson, 2000). Global surveillance is appropriate because WV is highly mobile and distributed by huge, macro-scale movements only observable from space. Generally,

Figure 4.10 *The invisible made visible: the dramatic fluidity of the upper troposphere is clearly demonstrated in this water vapour image (Meteosat, water vapour, 23 September 2003).*

weather satellites 'see' only the upper surface of the WV column (Figure 4.11) but some remote sensing systems can measure through the atmospheric column.

Water vapour images

Detecting features in grey-tone WV images is relatively easy because tonal contrasts are usually strong, features are large and homogeneous, patterns are distinctive and features persist in geographical space. As WV acts as a passive tracer, time-sequenced images are excellent for following meso-scale motions in the upper troposphere (e.g., wet air spiralling in tropical cyclones; dry intrusions behind cold fronts in mature mid-latitude depressions (see Section 2.4)).

Interpreting and explaining what is detected is quite a different matter. Effective image interpretation depends on understanding water vapour's role in the atmosphere.

Water vapour is:

- ubiquitous in the troposphere;
- distributed unevenly through space and time;

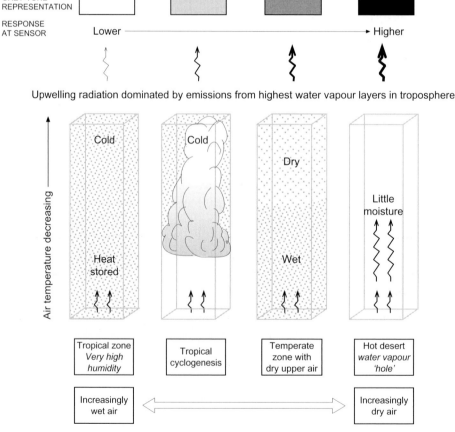

Figure 4.11 *Seeing the water vapour: representation of emitted energy in images. The sensor records infrared energy emitted through a column of the atmosphere but the signature from the upper water vapour layers (e.g., cloud-tops) is dominant. The brightest features are the coldest; the darkest features are dry (e.g., water vapour 'holes' over deserts).*

- strongly influenced by geographical location;
- highly mobile;
- a major heat source fuelling weather systems;
- concentrated in the lower troposphere (99% <5 km);
- abundant at the tropics, sparse in polar regions (500:1 ratio);
- found at high altitudes (>12 km) in the tropics, low levels at the poles (<8 km);
- found at very low concentrations over hot deserts (WV 'holes');
- sensitive to temperature (phase) change;
- highly responsive to aerosols;
- a powerful indicator of atmospheric stability/instability;
- a strong absorber of energy at various wavelengths;
- the major greenhouse gas.

The convention is to classify WV data in the reversed sequence of grey tones (Box 4.1) so that brighter tones indicate emissions from moist, but very cold, upper air (Figure 4.10). It is not just surface emissions that are detected, although these tend to be dominant, but the sum total of emissions (around 6.7 μm) propagated through the atmospheric column (Figure 4.11).

Brightness levels in WV images vary because:

- the atmosphere contains variable amounts of water vapour;
- air temperatures decrease with altitude;
- air temperatures vary with latitude and seasons;
- emissions originating from Earth's surfaces vary (e.g., low/ice, high/deserts);
- clouds contain water vapour;
- jet streams may be wet or dry;
- ascending air cools;
- descending air dries and warms;
- clouds are dynamic (variable/shifting brightness patterns).

Although darker tones generally indicate warmer, drier air, geographical location and circulation patterns must be considered. Vapour distribution is three-dimensional and the product of vertical and horizontal motions; interpretation and analysis is by no means easy.

Time-sequenced images are very helpful; as a general rule, if the image feature darkens, warming and drying is occurring, if it lightens, air is cooling and moisture content is increasing. However, it should be borne in mind that complex systems rarely obey simplistic 'rules'!

Darker tones may be associated with:

- subsiding air (e.g., dry slot in mid-latitude cyclogenesis);
- horizontal transfer of drier air (e.g., advected air);
- reduced cloudiness.

Lighter tones may be associated with:

- rising air (e.g., cyclogenesis; convection);
- advection of moist air (e.g., wet jet stream);
- increased cloudiness.

The rate of change and upper troposphere movements can be estimated – if distance and time factors are known. The evolution of cyclonic systems is clearly demonstrated in animated sequences.

The dark slot, behind a bright curving cloud, shrinks and changes position as the rapidly descending (drying) air is balanced by the vigorously rising (moist) air at the rising warm conveyor belts (Section 2.4).

Water vapour in clouds

Clouds contain water vapour and contribute to emissions around 6.7 μm. Gaps in cloud cover can be 'discovered', and air moisture content assessed, by inspecting a VIS image of the same scene. The textured, very bright clusters of cumulonimbus clouds are readily detected (Figure 4.10); less textured cloud types are more difficult to detect and identify.

A good deal is known about the behaviour of WV but scientists acknowledge that many uncertainties remain despite decades of observations using a plethora of techniques. Observational 'gaps' still exist over remote oceanic areas and in the middle atmosphere; extrapolating from data-rich to data-poor regions, and from 'local' to global scale, is fraught with difficulties. Space-based observations, always in conjunction with *in situ* observations, are the only practical way of regularly gathering WV data at a scale appropriate to understanding how the whole energy system works and, critically, how it is changing. Recent data indicate that WV concentrations have been increasing in the stratosphere: this observation remains unexplained (Schiller, 2001).

VISUALIZING HEAT: INFRARED IMAGES

As the human eye is insensitive to infrared wavelengths, heat released from the Earth is invisible unless it is extremely hot and emissions shift to visible (colour) wavelengths. However, we are able to measure and map the heat energy that leaves the Earth system from space (Figure 4.12). Infrared emissions are measured (day and night) by sensors tuned to respond to outgoing wavelengths between 3–5 μm and 8–14 μm transmitted through atmospheric windows (Figure 4.4).

Long-wave radiation, between 8 and 14 μm, is largely determined by an object's temperature (Box 4.5); wavelengths between 3 and 5 μm are both reflected and emitted during sunlit hours (only emissions

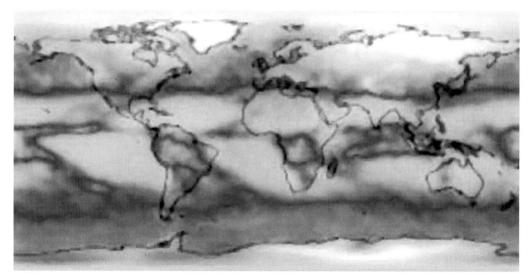

Figure 4.12 *Outgoing heat radiation. This shows all the emitted infrared energy (between 5 and 100 μm) escaping Earth's atmosphere (December 2002). The central light region indicates that most heat loss is from equatorial and desert regions. Polar areas release the least heat energy, shown 'light' only because the original is coloured! (NASA/Terra/CERES – see CERES Website for more information and animated false-colour images).*

Box 4.9 A bit of both around 3.7 μm

The behaviour of infrared energy changes as wavelengths lengthen between 3.0 and 4.0 μm; shorter waves are reflected, longer emitted. During sunlit hours, reflected infrared energy dominates the scene, swamping the emitted infrared, but at night only emissions will be recorded in the absence of solar radiation.

The data (e.g., **GOES Channel 2**) are represented in the conventional (reversed) way so that in night images the warmest objects (e.g., oceans) appear dark and cloud-tops white. Daytime scenes may appear as 'negatives' of night scenes, producing some strange-looking images, as highly reflective objects appear very dark (e.g., black ice and clouds).

Sunglint off calm, or even slightly ruffled, sea surfaces is often extremely distinctive in daytime images appearing as silvery grey, shiny patches because strong specular reflection dominates the scene.

occur at night). This 'awkward' spectral region (Box 4.9) is advantageous for meteorologists because certain phenomena, undetectable at other wavelengths, can be distinguished. Night images are particularly useful for detecting low warm cloud and fog but this depends on the warmth of the underlying surface. The difference may be very slight so image enhancement is needed to exaggerate (enhance) the contrast. **Difference images**, using two spectral bands (e.g., GOES Channel 4–Channel 3) are especially suitable for locating fog (Bader *et al.*, 1995: 447).

Thermal images: detection and interpretation

Thermal images can be confused with VIS images owing to the convention of reversing the grey-tone order (Box 4.1); the information they contain is, however, quite different. Grey tones correlate inversely with photon intensity (at the sensor) and are interpreted as brightness temperatures that are related to scene temperatures. Much can be inferred from the image simply by examining brightness patterns and boundary (edge) conditions; colour-coded images are extremely informative as more detail is revealed.

Brightness or **radiative temperatures** are not the same as actual temperatures – as measured by a thermometer. This means that visual interpretation, albeit extremely valuable, can only be qualitative. The actual, **kinetic temperature** can be derived for quantitative analyses. It is dependent on an object's emissivity, the sensor's properties, and measurement conditions.

Key points (refer to Figures 4.1 and 4.13):

- Brighter, lighter tones represent colder features; duller, darker tones represent warmer objects.
- Cloud-top temperature and height can be inferred – temperature generally decreases with height.
- The vertical organization of layered clouds can be inferred – higher, colder features appear brighter.
- Glaciated cloud-tops are brighter (colder) than non-glaciated, warmer clouds.
- Polluted clouds may absorb more energy and appear duller (warmer) than unpolluted clouds.
- Clouds associated with convection are relatively bright (e.g., cumulus).
- Less dynamic layer clouds appear duller (e.g., stratus).

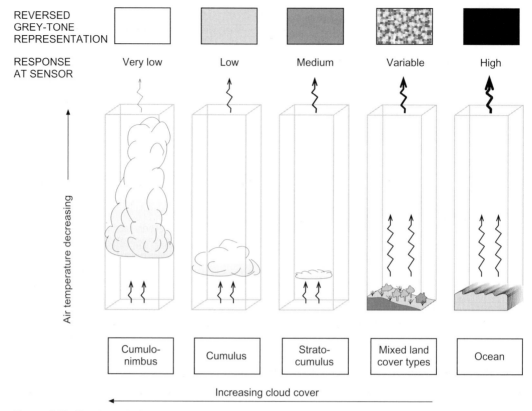

REVERSED GREY-TONE REPRESENTATION

RESPONSE AT SENSOR

Very low   Low   Medium   Variable   High

Air temperature decreasing

Cumulo-nimbus   Cumulus   Strato-cumulus   Mixed land cover types   Ocean

Increasing cloud cover

Figure 4.13 *Visualizing the heat: representation of emitted infrared in satellite images.*

- Clouds associated with strong convergence uplift are very bright (e.g., especially during cyclogenesis with strong vorticity).
- Grey 'smudges' may suggest low, thin, cloud cover. However, if air is close to dewpoint in humid tropical regions, this may be a false interpretation; the sky is likely to be cloud-free but super-saturated with moisture.
- Fog and ship tracks may be much the same temperature as the underlying surface in which case they will be 'invisible' as both surfaces will be emitting at similar wavelengths. In VIS images, fog will be 'seen' because of strong scattering of incident energy. This is particularly useful during anticyclonic periods as winter radiation (inland) fog and advection (sea) fog in spring and summer can be detected (also see Box 4.9).

It is important to remember that interpretations should not be based solely on spectral information; cloud shape, size, pattern, texture, transparency (optical thickness) and context need to considered, as well as comparison with VIS and WV images. The general 'rules' do not always apply.

4.6 THE SCENE BELOW: SEA, LAND AND ICE

Understanding weather from above depends on being able to 'read' the scene below the clouds since Earth's surface properties strongly influence upwelling radiances.

Water is the critical substance: it is ubiquitous and always in temporary storage in the hydrological system (i.e., the atmosphere, the cryosphere, geosphere and biosphere). Interpreting general boundary conditions, at the Earth–atmosphere interface, depends on understanding the behaviour of photons when they interact with the wet, dry and frozen surfaces of the oceans, land and ice masses.

OCEAN–ATMOSPHERE BOUNDARY

It is vital to observe ocean surfaces from space because huge quantities of energy and matter are exchanged at this critical boundary. Low reflectivity and high emissivity generally ensure that oceans appear as dark backgrounds to the bright whiteness of clouds in standard visible and infrared images. However, this is not always true as sea surfaces and near-surface layers are not homogeneous.

Visible images

Reflected/scattered energy is noticeable as lighter/brighter tones in oceanic scenes. Diffuse reflections occur from sea-bottom surfaces in clear shallow water (e.g., from submerged sandbars, bleached coral reefs, artificial structures) and from algal blooms and sediment plumes; in open water, ship exhausts scatter short-wave energy above the sea's surface. Identification of features is usually based on pattern recognition, location and spatial context. Differentiation of thin, low-lying clouds and dust plumes may be compromised by spectral 'pollution' especially from sunglint; specular reflections off calm water surfaces appear as very bright sheens in otherwise dark seas. Sunglint is easily recognized where constrained and configured by the coastline; clouds and dust can cross the land/sea boundary.

Identification of atmospheric phenomena is assisted by time-sequenced images as most marine features are less mobile than clouds and dust.

Thermal images and sea surface temperature (SST) maps

The brightness (recorded by sensors) and actual temperatures are similar because water behaves very like a theoretical blackbody emitting almost as much energy as it absorbs. Sea surface emissions are higher than most cloud-tops and land surface emissions at night and so appear darker in images.

Maps are constructed from derived **sea surface temperatures** (SST) that are visually dramatic and highly informative (Figure 4.14) and animated, archived data provide compelling evidence of shifting thermal patterns and temperature anomalies. Temperature anomalies identified in Arctic waters indicate

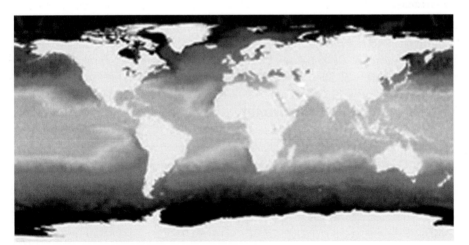

Figure 4.14 *Sea surface temperatures. Thermal state of the oceans as measured by the Moderate Resolution Imaging Spectroradiometer (MODIS), 2–9 June 2001. Cold waters are black. Grey tones represent progressively warmer water. White regions are the warmest (see MODIS Website for coloured versions).*

rapid melting of ice masses and rhythmic fluctuations in the Pacific's 'warm pool' correlate with El Niño/La Niña events (see Figure 6.7). This type of information is essential for modelling climate change and forecasting severe weather events. Significant links have been identified between thermal conditions in the tropical Pacific Ocean, extreme weather and hazardous, economically and environmentally disruptive events such as fires, floods and droughts.

Standard thermal infrared images often need to be enhanced and/or colour-coded to reveal subtle variations in ocean emissions undetectable in grey-tone products.

The oceans are so vast that it is difficult to believe that any biomass in the water could appreciably affect its thermal state or have any impact on our weather. But, as more insight is gained into the complex behaviour of natural systems, scientists acknowledge that local, apparently insignificant, changes may have unexpected cumulative (systemic) consequences. Systems (e.g., **SeaWiFS**) that are spectrally sensitive to organic pigments detect ocean colour and enable scientists to calculate biological energy consumption; this information is crucial to understanding the biochemistry of the atmosphere and the Earth's energy budget.

LAND–ATMOSPHERE BOUNDARY

Land surface cover is very mixed, often lacks spatial continuity and may vary over relatively short timescales (e.g., vegetation, snow cover) and, as a result, the behaviour of incident energy is hugely complex. Fortunately, it has been possible to build up a 'library' of **spectral responses** or **signatures** that characterize major and minor cover types (see remote sensing texts).

Visible images

Excluding ice-covered, permafrost and very wet regions, land is moderately reflective so it appears brighter than the oceans but duller than clouds in images. Water features and densely forested surfaces appear darker; deserts, semi-arid and desertified regions, deforested areas, urbanized and non-irrigated agricultural lands usually have higher albedos and appear lighter. The Earth's albedo has probably increased significantly as a result of land cover changes over thousands of years; it is impossible to know for certain because it has only been measured for about 30 years.

Thermal images

Land absorbs incident energy and releases it relatively quickly. Land surfaces generally appear duller (warmer) than cloud-tops and brighter (cooler) than oceans in winter (night) images. Very low-level clouds/fog may be thermally 'invisible' above land surfaces; detection is entirely dependent on the temperature and emissivity of different cover types.

CRYOSPHERE–ATMOSPHERE BOUNDARY

Boundary conditions here are critical as these systems are strongly coupled. Remote as it may seem, what happens in polar regions impacts the climate and weather of much of the world. Regular mapping and monitoring of changes in the **cryosphere** is only possible from space; scientists have no other means of collecting quantitative information at an appropriate scale and at regular intervals. Since data have been routinely collected from polar orbiting satellites, scientific understanding of boundary conditions has advanced significantly.

Visible images

Ice, snow (see Figure 5.25) and frozen ground (permafrost) strongly reflect incident energy appearing very bright/bright in VIS images. Melting surfaces appear much duller as the released water absorbs rather than reflects energy: absorption promotes further melting and reduces reflection. Particle-laden surfaces similarly change the way incident energy behaves (Box 4.10).

Thermal images

In conventional thermal images these cold surfaces appear bright; melting and debris-loaded surfaces are warmer and appear duller (Box 4.10). Clouds are perceived as brighter against melting surfaces as the contrast is enhanced.

Box 4.10 Snow, ice and frozen ground

Cryospheric surfaces are highly reflective because the molecular properties of water change below freezing point. During the **phase change**, molecules bond together more tightly inhibiting transmission and absorption but encouraging strong reflection. During melting, molecular coherence (bonding) decreases, absorption increases as liquid is released and reflection decreases. The physical state, purity and age of ice affect its albedo; older (glacier) ice becomes less reflective and more absorbent as structural coherence decreases and surfaces accumulate particulate material. An ice-free world would be a very much wetter and warmer place.

Fresh snow appears very bright because reflection is promoted by the highly organized molecular structure of snowflakes; strong bonding inhibits absorption. The shape, size and liquid content of snowflakes, and the depth of the snow layer, will determine how much energy is reflected. Larger, well-organized flakes reflect more and absorb less; melting triggers a decrease in molecular coherence, absorption increases and snow appears dull. Thin layers of snow have relatively low albedos compared with thick snow (>13 cm) so, in a warming world, less snow will promote further atmospheric warming – a positive feedback scenario.

Frozen ground (permafrost) plays a significant role in the Earth's energy budget because it is widespread, especially in the northern hemisphere, seasonally highly variable and stores a huge amount of frozen water.

Frozen surfaces have temperatures well above 0 K, so emit energy at very long wavelengths; crystal size, water content and internal temperatures all affect emissions. If global temperatures continue to rise, more energy will be absorbed and more emitted in a wetter world; cloud cover will increase, trapping more heat and further enhancing the greenhouse effect.

In cloud-free images the brightness of ice and snow indicates its freshness; duller surfaces indicate decaying and/or particle-rich surfaces. Any reduction in brightness over a long time period would indicate warming; increasingly bright scenes would suggest global cooling. In reality the system is much more complex than this suggests because of feedback processes.

Sea ice

The spatial and temporal characteristics of sea ice are monitored from space. This information is crucial as the amount of sea ice correlates with rising global temperatures and affects the spatial characteristics of cold and warm ocean currents. Swarms of bright sea ice are easily detected against a dark sea; animated time-sequenced images are especially useful for monitoring movement.

Current and archived Earth Observation data and field observations indicate that ice masses and frozen ground phenomena are melting at a rate faster than previously recorded. Melting permafrost and shrinking ice masses absorb more energy triggering positive feedback by warming the lower atmosphere and promoting further melting.

Equally tightly coupled is the cryosphere and oceanic system. The release of extra quantities of cold water in polar regions perturbs ocean currents and has a knock-on effect on the weather. Seasonal changes in ice cover, ice calving and the state of permafrost regions is monitored from space using visible, infrared and radar techniques. If the climate is warming, land surfaces released from the grip of very low temperatures will be extremely unstable owing to high saturation levels and the absence of vegetation. More rivers will discharge huge quantities of cold water and sediment into the oceans. The faster the ice and permafrost melts, the more impact this will have on the Earth's energy budget and other systems: exactly how this will impact climate and weather is unknown.

SUMMARY

Scientists have skilfully constructed elegant but, nevertheless, simplistic models of the Earth's energy system that bear little resemblance to the real thing. Scientists urgently need to describe the atmospheric system accurately, at a global scale, and to measure and monitor natural and human-induced perturbations in the Earth's energy budget. Most scientists acknowledge that, while a good deal is known about small-scale, isolated mechanisms, very little is understood about interactions between systems and feedback processes: the key word that pervades scientific publications is 'uncertainty'! It is in this context that remote sensing has 'come of age' because it routinely provides quantitative data essential for, amongst other things, more reliable climate and Earth systems modelling that is helping scientists better understand the complexity of the system. Nevertheless, a word of caution: Earth Observation science is still in its infancy – a great deal has yet to be learnt – but we are moving in the right direction.

We find only what we look for and the more we look, the more we see; the more we see, the better we know where to look. Space-based technologies enable scientists to observe previously unobservable phenomena and, in synergy with *in situ* measurements, enable us to better understand our world, our climate and the weather.

REFERENCES AND GENERAL READING

Bader, M.J., Forbes, G.S., Grant, J.R., Lilley, R.B.E. and Waters, A.J. 1995: *Images in weather forecasting.* Cambridge: Cambridge University Press.

Burroughs, W.T. 1991: *Watching the world's weather.* Cambridge: Cambridge University Press.

Burroughs, W.T. (ed.) 2003: *Climate: into the 21st century.* Cambridge: Cambridge University Press.

Browning, K.A. 1994: Life cycle of a frontal cyclone. *Meteorological Applications* 1, 233–35.

Browning, K.A. and Roberts, N.M. 1994: Use of satellite imagery to diagnose events leading to frontal thunderstorms: part 1 of a case study. *Meteorological Applications* 1 (4), 303–10.

Brugge, R. and Stuttard, M. 2003a: Back to basics: from Sputnik to Envisat, and beyond: the use of satellite measurements in weather forecasting and research: Part 1 – a history of weather. *Weather* 58, 107–12.

Campbell, J.B. 1996: *Introduction to remote sensing.* 2nd edition. London: Taylor and Francis Ltd.

Chen, J.Y., Carlson, B.E. and Del Genio, A.D. 2002: Evidence for strengthening of the tropical general circulation in the 1990s. *Science* 295, 838–41.

DeGrand, J.Q., Carleton, A.M., Travis, D.J. and Lamb, P.J. 2000: A satellite-based climatic description of jet aircraft contrails and associations with atmospheric conditions, 1977–79. *Journal of Applied Meteorology* 39, 1434–59.

Elsberry, R.L. and Velden, C. 2003: A survey of tropical cyclone forecast centres – uses and needs of satellite data. *Bulletin of the World Meteorological Organisation* 53, 258–64.

Fabry, F. and Zawadzki, I. 2002: New observational technologies: scientific and societal impacts. In Pearce, R.P. (ed.) *Meteorology at the millennium.* London: Academic Press.

Gibson, P.J. 2000: *Introductory remote sensing: principles and concepts.* London: Routledge.

Gurney, R.J., Foster, J.L. and Parkinson, C.L. 1993: *Atlas of satellite observations related to global change.* Cambridge: Cambridge University Press.

Harries, J.E. 2002: The impact of satellite observations of the Earth's spectrum on climate research. In Pearce, R.P. (ed.) *Meteorology at the millennium.* London: Academic Press.

Hastings, D. 1998: Advanced very high resolution radiometer (AVHRR) overview. NOAA national data centres web site Washington DC. http://www.ngdc.noaa.gov/seg/globsys/avhrr.shtml

Horvath, A. and Davies, R. 2001: Simultaneous retrieval of cloud motion and height from polar-orbiter multiangle measurements. *Geophysical Research Letters* 28, 2915–18.

Hutchinson, K.D. 2002: The retreival of cloud base heights from MODIS and three-dimensional cloud fields from NASA's EOS Aqua mission. *International Journal of Remote Sensing* 23, 5249–65.

Kemp, D.D. 1994: *Global environmental issues: a climatological approach.* 2nd edition. London: Routledge.

King, M.D., Kaufman, Y.J., Tanré, D. and Nakajima, T. 1999: Remote sensing of tropospheric aerosols from space: past present and future. *Bulletin of the American Meteorological Society* 80, 2229–59.

Marquart, S., Ponater, M., Mager, F. and Sausen, R. 2003: Future development of contrail cover, optical depth, and radiative forcing: impacts of increasing air traffic and climate change. *Journal of Climate* 16, 2890–904.

Meyer, R., Mannstein, H., Meerkotter, R., Schumann, U. and Wendling, P. 2002: Regional radiative forcing by line-shaped contrails derived from satellite data. *Journal of Geophysical Research – Atmospheres* 107, 4104.

Minnis, P., Schumann, U., Doelling, D.R., Gierens, K.M. and Fahey, D.W. 1999: Global distribution of contrail radiative forcing. *Geophysical Research Letters* 26 (13), 1853–56. Also see http://ams.confex.com/ams/13ac10av/10ARAM/abstracts/40538.htm

Myhre, G. and Myhre, A. 2003: Uncertainties in radiative forcing due to surface albedo changes caused by land-use changes. *Journal of Climate* 16, 1511–24.

Nober, F.J., Graf, H.F. and Rosenfeld, D. 2003: Sensitivity of the global circulation to the suppression of precipitation by anthropogenic aerosols. *Global and Planetary Change* 37, 57–80.

Parker, C.L. 2003: Aqua: an Earth-observing satellite mission to examine water and other climate variables. *IEEE Transactions on Geoscience and Remote Sensing* 41, 173–83.

Rast, M., Bezy, J.L. and Bruzzi, S. 1999: The ESA medium resolution imaging spectrometer MERIS – a review of the instrument and its mission. *International Journal of Remote Sensing* 20, 1681–702.

Schiller, C. 2001: What mechanisms do control the water vapour distribution in the lower stratosphere? 15th ESA Symposium on European Rocket and Balloon Program and Related Research. *Proceedings ESA Special Publication* 471, 63–66.

Stenger, R. 2002: 9/11 study: air traffic affects climate. http://edition.cnn.com/2002/TECH/science/08/07/contrails.climate/index.html

Toon, O.B. 2000: How pollution suppresses rain. *Science* 287, 1763–64.

Travis, D.J., Carleton, A.M. and Lawitsen, R.G. 2002: Contrails reduce daily temperature range. *Nature* 418, 601.

Wielicki, B.A., Wong, T.M., Allan, R.P., Slingo, A., Kiehl, J.T., Soden, B.J., Gordon, C.T., Miller, A.J., Yang, S.K., Randall, D.A., Robertson, F., Susskind, J. and Jacobowitz, H. 2002: Evidence for large decadal variability in the tropical mean radiative energy budget. *Science* 295, 838–44.

Wentz, F.J. and Schabel, M. 2000: Precise climate monitoring using complementary satellite data sets. *Nature* 403, 414–16.

5

Climates of the mid-latitudes

In this chapter, the global distribution and characteristics of mid-latitude climates are considered before examining the climatic features of Europe, the Mediterranean Basin and the British Isles. The climate and weather of these regions is governed by the location and vigour of the prevailing mid-latitude westerly winds. The focus then changes from these essentially maritime, west coast climates to those of North America. While much of this continent lies within the mid-latitudes, it is a region in which continental influences are more developed and one in which meridional (north–south) exchanges of air have a more profound influence on weather and climate compared with Europe.

5.1   THE DISTRIBUTION OF MID-LATITUDE CLIMATES

Mid-latitude climates occur poleward of the sub-tropical high pressure systems (30–35°N/S) and equatorward of the mid-latitude low pressure systems (typically near 60°N/S). They are found in every continent with the exception of Antarctica. Their extent is influenced by the distribution of land and sea. In the southern hemisphere only small areas of land reach the mid-latitudes; the south of Chile and Argentina, Tasmania, the far south of Australia (Victoria state and Tasmania) and much of New Zealand. In the northern hemisphere much of the USA, southern and central Canada, most of Europe and parts of Asia lie in the mid-latitudes.

EFFECT OF SEASON
As shown in Chapter 2, the mid-latitude westerly winds are driven by temperature gradients. These gradients (or contrasts) peak in winter, resulting in a stronger and more dominant westerly circulation of air. These temperature contrasts provide the atmospheric fuel for the development of low pressure systems. An important feature of mid-latitude climates is the seasonal migration of pressure systems to a lower latitude in winter, a response to the equatorward movement of the temperature gradients, a change triggered by the formation and spread of sea ice.

By late spring and early summer, the temperature gradients have weakened, reducing the vigour of mid-latitude depressions. This leads to a seasonal rainfall minimum in many areas. A blocked weather type is more likely at this season as the westerly winds become lighter and more meandering in their circumpolar motion.

EFFECTS OF LONGITUDE AND LATITUDE
As westerly winds pass over oceans, moisture is evaporated into the atmosphere. Much of this moisture is released as air rises over the often hilly west-facing coasts. Examples include the Western Cordillera of North America, the mountains of Ireland, the British Isles and Norway and the extreme southern Andes

in Chile. As the westerly winds penetrate the largest land-masses they become drier and the weather and climate become more continental in nature in central and eastern parts of North America and Asia.

The effect of latitude is related to the varying seasonal influence of areas of high and low pressure. As latitude decreases, the influence of high pressure develops at the expense of low. This can be demonstrated by gradual transition to a summer dry season on the equatorward margins of each mid-latitude climate zone in areas having a Mediterranean type climate. Rainfall seasonality is caused by the poleward retreat of the westerlies in the summer and the arrival of the dry, descending air of the sub-tropical high pressure. This high pressure is a semi-permanent feature of the summer months. It is found on the western seaboards of all continents at the intersection of mid-latitude and sub-tropical climates (typically around 35°N/S); in California, central Chile, around the Azores, the extreme south of Africa and in the south of Western and Southern Australia.

5.2 THE CLIMATE OF EUROPE

Europe combines two transitions; a latitudinal transition from sub-arctic to sub-tropical and a longitudinal transition from maritime to continental. The north/south transition is sharpened by mountain ranges acting as climatic barriers whereas the eastward penetration of sea in places extends the maritime influence eastwards (Figure 5.1).

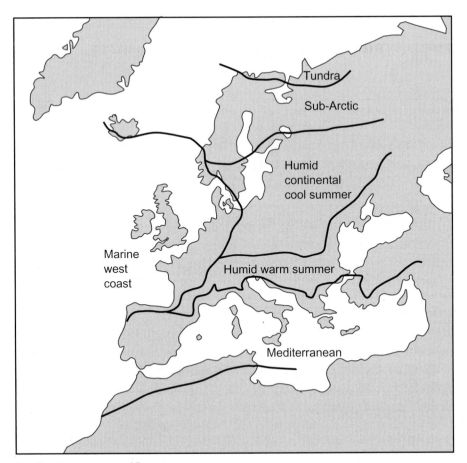

Figure 5.1 The climatic regions of Europe.

GEOGRAPHICAL CONTROLS ON THE CLIMATE OF EUROPE

Lying principally within the mid-latitudes on the western side of a large land-mass, the effects of maritime westerly winds can spread inland with little hindrance. This effect is enhanced by the eastward penetration of sea in both northern Europe (the Baltic) and the south (the Mediterranean). Europe is unusual in having an extensive west–east mountain range. This extends from the Cantabrian Mountains of northern Spain through the Alps to the Caucasus Mountains. These mountains lie at about 42°–45°N, corresponding to the climatic boundary between the mid-latitudes (with rain in all seasons) to the north and the Mediterranean climate to the south.

Central southern Europe around the Alps is the only part of the world where humid warm summer climates are found between west coast mid-latitude and Mediterranean climates (exemplified by Geneva – Figure 5.2). This is due to the eastward extension of a moisture source – the Mediterranean Sea – and the presence of high ground. Mountain barriers emphasize geographical contrasts in weather and climate by blocking the passage of air masses and by modifying the character of air when it passes over high ground.

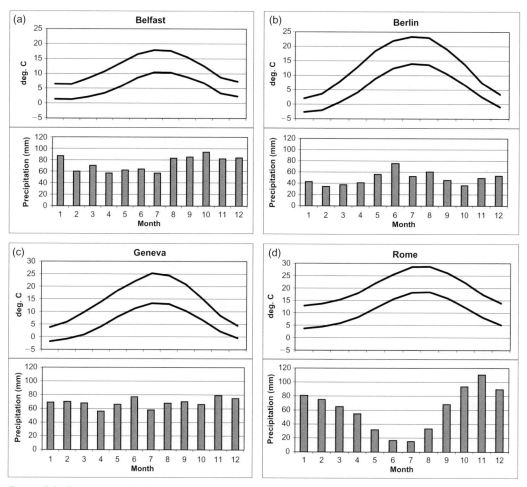

Figure 5.2 *Climate graphs for four contrasting areas: (a) Belfast (west coast maritime); (b) Berlin (continental northern Europe); (c) Geneva (humid warm summer); (d) Rome (Mediterranean).*

Areas to the lee of high ground are often warmed by the föhn effect, the warm downslope wind characteristic of the Alps and many other mountainous areas in Europe (Section 1.8 and Box 5.1).

Several other geographical controls operate at a variety of scales:

- **latitude** has a direct influence upon temperature by determining the receipt of solar radiation. Latitude also determines the exposure to different aspects of the mid-latitude circulation of air as contrasting air masses converge around areas of low pressure;
- **ocean currents** influence the character of air masses that originate over the eastern North Atlantic, setting up the contrast between a maritime west and a more continental east that is such a well-known feature of climate and weather in Europe;
- as in other parts of the world, the **local geographical setting** determines the character of meso and micro-climate – the interplay of soil type, local orography, local winds and patterns of shelter and exposure.

SEASONAL VARIATIONS IN TEMPERATURE AND PRECIPITATION

This overview of the climate of Europe examines the changing seasonal patterns of temperature and precipitation, evaluating the role of the geographical controls on climate.

Box 5.1 The föhn effect in Europe – examples from Spain, Poland and Iceland

The föhn effect takes its name from the local southerly wind descending from the Alps in south Germany. Forced descent warms the air with quite spectacular effects on local temperature when the air is stable, as is most typical in southerly winds. This explains why it is more common in Munich than Milan, south of the Alps. It is renowned as a cause of rapid snowmelt in the Alps, sometimes contributing to avalanche risk.

Föhn winds can, however, occur in any wind direction. If a stable southerly airstream covers Spain, places such as Santander on the north coast may become unseasonably warm as the air descends rapidly from the Cantabrian Mountains. Biarritz, in southwest France, possibly owes its mild reputation to descent of air in the lee of the Pyrenees. Other favoured locations for a southerly föhn include Algiers and the north coast of Sicily. In Poland they are sufficiently frequent to raise the mean temperature by about 1 deg. C and to reduce the average relative humidity by about 10% in the lee of the Carpathian mountains (Ustrnul, 1992; Quaile, 2001).

While the föhn effect can operate in fine weather, it is particularly effective when the air ascending the high ground is saturated and thereby only cools at the saturated adiabatic lapse rate. The effect can therefore be especially pronounced in northwest Europe in winter when the air is moist. Eden (1995) identified one occasion when the warmest place in Europe was at Dalatangi: this exotic-sounding location is actually in eastern Iceland. On the day in question, 18 January 1992, a mild, stable southwesterly breeze allowed the temperature to reach 18°C, a new Icelandic record for January. Dalatangi also achieved a new record November temperature for Iceland on 10 November 1999 when it reached 22.7°C (the previous record at this site being 19.7°C).

Temperature

Winter temperatures

The pattern of mean temperature in January results from the warming influence of the North Atlantic and the presence of the North Atlantic Drift, these influences being enhanced by the frequency of southwesterly winds. There is a marked distortion of poleward cooling as isotherms are aligned north–south over northern and central Europe (Figure 5.3).

Mean January temperatures are above 0°C along the Norwegian coast as far as the Arctic Circle and the Norwegian Alps provide a sharp boundary between maritime and continental air. Further south across Denmark and Germany in the absence of high ground, the eastward cooling is more progressive. The isotherms run west–east over southern Europe, a product of the temperature gradient between the relatively warm Mediterranean Sea to the south and high ground to the north.

Summer temperatures

In July, temperatures across Europe increase southwards and southeastwards, reflecting a dominant influence of latitude. The coolest area of Europe is now the northwestern seaboard which is now cooled by maritime air and sea currents. Mean July temperatures here average less than 15°C. The 15°C isotherm actually crosses the Arctic Circle just north of the Gulf of Bothnia.

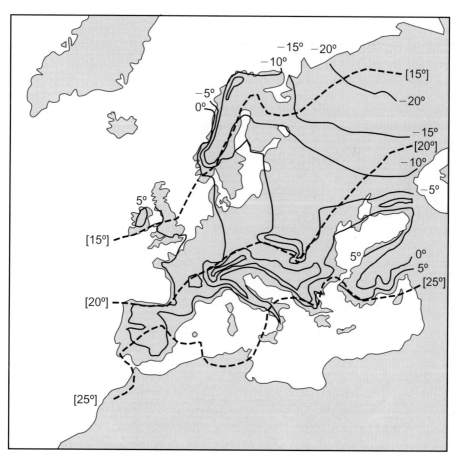

Figure 5.3 *Mean temperature in Europe in January (continuous isotherms) and July (dashed isotherms and square brackets).*

The Mediterranean Sea fails to provide the moderating influence on summer temperature that it achieves in winter. Because it is an almost land-locked body of water (with little movement of water through the Straits of Gibraltar) it warms up steadily in summer, diminishing the potential for sea-breezes.

Characteristics of a maritime climate

The average annual range of mean monthly temperatures varies from under 10 deg. C on the western seaboard of Europe to over 25 deg. C from Finland and Russia eastwards, reaching more than 35 deg. C east of the Ural Mountains. A feature of maritime air masses is the tendency for temperature to drop rapidly with height. The resulting high lapse rates have a marked effect on vegetation, especially in the maritime uplands of the British Isles and Norway. In southwest England the growing season decreases on average by 9 days for every 30 m increase in altitude; the corresponding figures for northern Britain and central Europe are 5 days and 2 days, respectively.

The length of the growing season in a maritime climate is especially sensitive to temperature changes. The growing season is usually defined as the duration over which the mean temperature exceeds 6°C. Because it takes longer to warm by 1 deg. C in a maritime climate in spring (compare Belfast and Berlin in Figure 5.2), an overall drop (or rise) in temperature shortens (or lengthens) the growing season by many more weeks.

Precipitation

Since most of Europe is situated in the zone of mid-latitude westerly winds between the average positions of the Icelandic Low and the Azores High, the wettest parts of Europe lie on the north-westernmost mountains where moist, maritime air is forced to ascend the west-facing slopes. Annual totals widely exceed 1500 mm over four mountainous regions, namely:

- *northwest Britain and the Norwegian Alps.* High ground is aligned north–south, enhancing contrasts between exposed, wet western slopes experiencing orographic enhancement of rain in westerly winds and drier eastern slopes. A sharp rain-shadow occurs in western Sweden in the lee of the Norwegian Alps where the annual average is less than 400 mm around 11°E/62.5°N;
- *the Swiss, Austrian and Yugoslavian Alps.* High annual totals characterize the Alps from Switzerland, through southern Austria, western Slovenia and Montenegro. In addition to orographic uplift, the mountains encourage thunderstorms in the summer half-year as a result of high ground acting as a heat source;
- *northwest Spain.* Northwest Spain and northern Portugal experience heavy frontal rain when warm, moisture-laden air rises over the hilly coasts – rainfall varies from over 1500 mm over coastal mountains to under 400 mm inland;
- *the eastern Black Sea.* The Caucasus Mountains rise sharply within 100 km of the eastern shores of the Black Sea. Annual rainfall varies from 2400 mm here to under 400 mm on the rain-shadow area to the northeast. The mountainous north coast of Turkey is also often very wet.

The driest area of Europe is in southeast Spain, where annual rainfall averages less than 250 mm in Almeria. The problems associated with water scarcity in this arid region are discussed in pp. 138–9. Rainfall averages less than 400 mm per year over wide areas of inland Spain and Turkey, illustrating the dryness of continental climates, despite the high altitudes found in both regions. However, while totals of under 600 mm are widespread across low-lying areas of northern Europe, the tendency of rain to be evenly distributed over time, and the usually modest evapotranspiration rates here, maximize the effectiveness of rainfall in maintaining soil moisture.

Winter precipitation

Winter precipitation is dominated by the wetness of coastal areas and the dryness of the continental interior of central and northeastern Europe. It is controlled by the vigour and extent of the mid-latitude westerly winds and their associated baroclinic zones. The wettest parts of Europe in winter are over the west-facing slopes of the Scottish Highlands and the Norwegian Alps (January average over 300 mm). Away from upland areas and western coasts, monthly rainfall is usually less than 50 mm in winter (this contrast is illustrated by Belfast and Berlin in Figure 5.2). The dominance of westerly winds in winter means that rain-shadow locations to the east of high ground are prominent. Heavy rain is also experienced across the Mediterranean, especially around west-facing mountains – over 250 mm in January in northern Portugal, on the Yugoslavian Alps and more locally in southern Italy and southern Turkey.

Summer precipitation

The main change in summer is the reduction in rainfall across the Mediterranean in response to the northward shift of mean depression tracks. The reduced vigour of frontal systems and Atlantic depressions also means that the contribution of frontal rainfall is lower than in winter. Summer rainfall over northwestern coasts and mountains can be less than half winter levels. Further east, convectional rainfall reaches peak intensity as rising temperatures encourage thunderstorms (Holt et al., 2001). These storms can be triggered by the Pyrenees and the Alps acting as high-level heat sources, creating a contrast between high summer rainfall totals here and the dry Mediterranean Basin to the south (e.g., Geneva in Figure 5.2).

ACCOUNTING FOR VARIATIONS OVER TIME – THE NORTH ATLANTIC OSCILLATION

The North Atlantic Oscillation describes how the atmospheric circulation across Europe and the northeast Atlantic varies over different timescales (Box 5.2). It is a large-scale influence on temperature, winds and precipitation patterns across the North Atlantic and Europe.

Box 5.2 Defining the North Atlantic Oscillation

The North Atlantic Oscillation (NAO) is an oscillation of the air pressure gradient between the Icelandic Low and the Azores High (Perry, 2000a). This relates to north–south shifts in the tracks of depressions, the intensity of depressions and changes in the vigour of mid-latitude westerly winds.

It is measured by the North Atlantic Oscillation Index (NAOI), quantifying the difference in air pressure between Iceland and a location representative of the Azores High (either in the Azores itself or in Portugal). If the difference is above average the NAO is in positive mode, implying a strengthened westerly circulation arising from a deepened Icelandic Low and/or an intensified Azores High. If it is below average, the NAO is in negative mode, implying a weakened circulation (however, the actual air pressure over the Azores will not necessarily be lower than in Iceland).

Positive mode NAO events are associated with below average temperatures over Greenland and Iceland and above average temperatures over continental Europe (a cold sea, warm land situation). Although the causes of variations in the NAO are not fully understood, it is known that much of the variation is driven by the pattern of sea surface temperature and may have an origin in changes in tropical sea temperatures (Rodwell et al., 1999).

Table 5.1 The characteristics of the atmospheric circulation during contrasting NAO events

	NAO positive mode	NAO negative mode
Upper airflow type	Progressive	Blocked
Index cycle characteristic	High index	Low index
Size of Rossby waves	Small	Large
Air pressure and temperature anomaly		
Iceland	Below average	Above average
Southern Europe	Above average	Below average

Variations in the North Atlantic Oscillation

The relationship between the NAO and the atmospheric circulation is shown in Table 5.1.

The positive mode of the NAO (Figure 2.3(a)) in winter usually allows mild Atlantic air to penetrate eastwards into Europe. This results in temperatures across Scandinavia and Russia being more than 5°C above average in places. Southern Europe may experience long dry spells – even in winter – as the Azores High will tend to cover much of the Mediterranean region. In summer a similar high index weather situation will bring cool and unsettled weather to northern Europe.

The weather associated with the negative mode (Figure 2.3(b)) depends upon the position of ridges and troughs in the upper circulation. This is often related to thermal influences. For example, if Europe is cold and the North Atlantic warmer than average, this will encourage a trough to develop over central or even southern Europe.

PROGRESSIVE WEATHER TYPES

Progressive weather types bring invigorated mid-latitude storms to northern Europe, with an increased risk of flooding. This section examines the characteristics of the severest storms and depressions that have affected Europe over recent decades.

Storms

Winters of the late 1980s and much of the 1990s were characterized by several notably stormy periods and a high frequency of gales (Palutikof et al., 1997), consistent with a rise in the winter NAOI. The effects of such storms are not simply the product of one-off events but often represent the culmination of several weeks of unsettled, windy weather. Namias (1987) has noted how progressive situations become self-sustaining over a period of weeks. The tendency for families of deep depressions to take similar routes is of significance for shipping because wave heights are often the product of more than one storm system (Burroughs and Lynagh, 1999).

In December 1986 a depression deepened explosively as it passed over the western North Atlantic, reaching a central pressure of 916 mbar between south Greenland and Iceland on the 15th (Burt, 1987). At the time, this was the deepest North Atlantic depression on record (it was surpassed by that of January 1993). This was a good example of a depression that formed explosively in a strong temperature gradient – North Atlantic sea surface temperatures were mostly below average north of 40°N (reaching an anomaly of more than −3.0 deg. C off Newfoundland and west of Iceland) and above average to the south of this latitude (Namias, 1987). The effects of the resulting baroclinicity were amplified by the resulting strong jet stream.

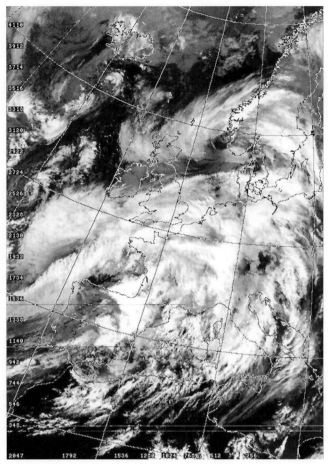

Figure 5.4 *The Great Storm of October 1987; NOAA AVHRR thermal infrared image for 15:11 h 15 October 1987, showing a cloud head to the southwest of Ireland as the storm developed.*

Perhaps the most notorious storm of the twentieth century in northwest Europe was that of 15 October 1987 (Figure 5.4). Though falsely called a 'hurricane' by the media, it actually had its origins in Hurricane (tropical cyclone) Floyd in the Caribbean one week earlier. It is not unusual for a mid-latitude depression to obtain some of its initial energy from the outburst of warm air from the top of a tropical cyclone, although the explosive cyclogenesis occurred mostly because of the warmth of the Bay of Biscay. This late development unlocked perhaps the most widely publicized weather forecasting challenge of the late twentieth century. However, the storm was unusual more for its low latitude (and hence impact) rather than its severity.

The winter of 1989–90 was one of the most progressive of the twentieth century. The most notable storm occurred on 25 January 1990 – the 'Burns' Day Storm' (a satellite image is shown on the front cover). The storm tracked right across the British Isles and affected all areas, unlike that of October 1987. Far more people actually had direct experience of the storm because it reached a peak during the working day. However, the impact upon trees (and hence power lines) was much reduced because deciduous trees were not in leaf and were a less vulnerable target.

Three years later the North Atlantic experienced one of the most progressive weather situations on record and the deepest depression yet measured. Strong winds in early January 1993 led to the sinking of the *Braer* supertanker off Shetland. One storm surpassed that of 1987 for deepness – whilst

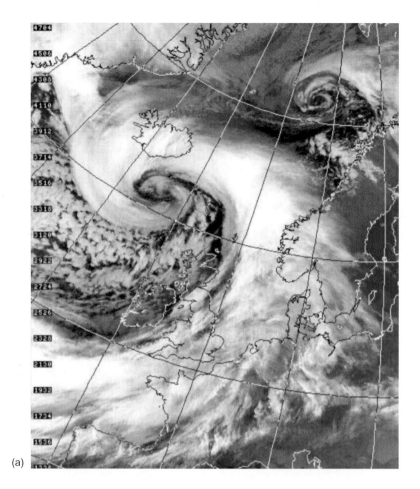

(a)

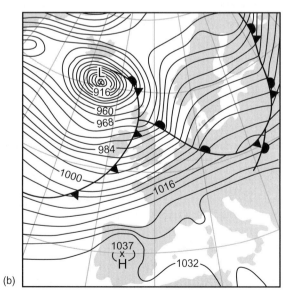

(b)

Figure 5.5 *The storms of January 1993. (a) NOAA AVHRR infrared satellite image for 14:39 h 10 January 1993, and (b) surface synoptic weather chart for 12:00 h.*

the central pressure fell to about 912 mbar south of Iceland on 10 January, pressure remained high over southern Europe, exceeding 1035 mbar over Spain; a pressure gradient of 119 mbar, indicative of an exceptionally high NAOI value. The exact central pressure will never be known; a floating buoy was located close to the depression centre but this was not designed to report air pressure below 920 mbar (Burt, 1993), a reminder of the fact that the value of observations is greatest when the operating conditions are at their most severe.

The scale of the 1993 storm can be appreciated from the fact that surface winds of gale force were reported throughout the North Atlantic and on the western seaboard of Europe from Spain to Norway. A submersible platform north of Scotland recorded a 'heave' of 11 m (Burroughs and Lynagh, 1999). It was a product of an exceptionally active jet stream, crossing the Atlantic with a core speed of 240 knots (McCallum and Grahame, 1993). The highest wind gust at the surface was 105 knots at North Rona, north of Scotland. Despite the rate of development, the storm was generally well captured by the forecast models. It was followed by a blast of very cold and unstable polar maritime air carrying a feed of heavy wintry showers towards Scotland (Figure 5.5).

The Christmas periods of 1997 and 1999 were marked by stormy weather in western Europe. Northern Britain experienced a memorable storm on Christmas Eve 1997 that threw many people's Christmas travel plans into chaos; it was also an example of how difficult it sometimes is to forecast explosive cyclogenesis. As the depression moved northeast towards Ireland on the 23rd, satellite imagery showed clear 'cloud head' and 'dry intrusion' structures in the cloud (Figure 5.6), precursors to explosive cyclogenesis (Young and Grahame, 1999).

Further evidence of storm development came from a drifting buoy that recorded a sharp reduction in air pressure in the eastern North Atlantic. Young and Grahame (1999) noted that this might have been discarded as an apparently incorrect observation had it not been for the visual guidance of the satellite imagery. Indeed, forecasters then introduced 'bogus' observations into the forecast model in order to enable it to capture the pressure reduction. In forecasting terms this was an interesting example of the symbiotic relationship between traditional instrumental observations (albeit on floating buoys!) and remote sensing. By Christmas Eve the storm culminated in wind gusts to 96 knots over North Wales. Transport and electricity supplies were badly disrupted in northern and western Britain and many people faced an anxious Christmas holiday without power.

December 1999 was notable for three highly vigorous depressions that crossed Europe at successively lower latitudes. In all, around 130 lives were lost in the resulting storm damage, which was particularly severe in France. The first storm (*Anatol*) arrived on the 3rd and was the worst storm in Denmark of the twentieth century, with a record storm surge. Winds speeds over the North Sea reached 97 knots (Ulbrich et al., 2001). *Anatol* was fed by a tongue of warm air to the south of Newfoundland interacting with cold air in an upper trough around Iceland. Continuing temperature contrasts allowed it to remain deepening for longer than usual – the lowest pressure was 953 mbar over Denmark.

The second storm (*Lothar*) crossed northern Europe on Boxing Day, tracking from northwest France (where explosive cyclogenesis was triggered) to Germany. Wind speeds were exceptional around Paris where extensive damage was caused to buildings and trees. The peak gust at Paris (Orly) Airport was 97 knots, an exceptional speed for a major European city. The vigour of the westerly airflow at the time was demonstrated by the speed of the depression – it moved eastwards at a speed of 65 knots powered by a jet stream core at about 50°N moving at about 200 knots. Another deep depression followed on the 27th (*Martin*) but took a slightly more southerly track (see satellite image showing 'cloud head' in Figure 2.16), resulting in flooding along the French Atlantic coast and severe damage to the forests of central and southern France (Pearce et al., 2001).

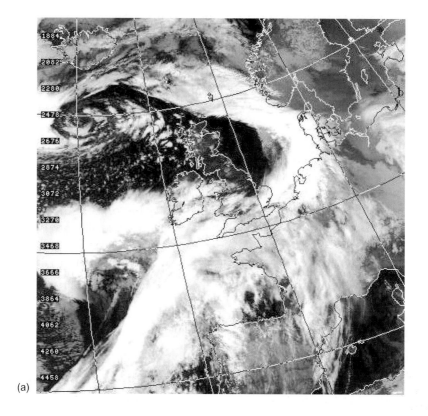

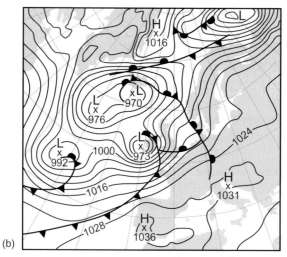

Figure 5.6 The Christmas Eve storm, 24 December 1997. (a) NOAA AVHRR thermal infrared image for 03:45 h; and (b) surface synoptic chart for 12:00 h. The storm was associated with explosive cyclogenesis of a secondary depression moving northeast towards the British Isles. The curl of cloud west of Ireland is a well developed 'cloud head', with a 'dry slot' of clear skies to the south. The solid white tone here indicates strong uplift and cold cloud-tops whereas the ragged or uneven appearance of the cloud south of Iceland indicates a lack of new cloud development around the parent depression centre.

WEATHER ASSOCIATED WITH BLOCKED WEATHER TYPES

Blocking of the North Atlantic westerlies promotes a more extreme seasonal temperature regime across Europe as oceanic influences are weakened. Blocking is usually associated with a large-scale diversion of rain-bearing depressions and fronts. The effect upon precipitation is usually subsidiary to that upon temperature, though the presence of a blocking high pressure over Europe will obviously tend to be associated with dry weather.

Winter cold

The winters of 1946–47 and 1962–63 plunged most of central and northern Europe into sustained freezing weather caused by persistent blocking of the mid-latitude westerlies (Figure 5.7). Much of Britain had continuous snow-cover in February 1947 with depths reaching 1.35 m in County Durham, UK, on the 19th. The Central England temperature was −1.9°C. The winter of 1962–63 was even colder, January being the coldest month in Central England (−2.1°C) since January 1814 (−2.9°C). Though there were heavy falls of snow, notably in southwest England, there was generally less snow than in 1947.

Europe experienced a number of cold spells in the 1980s and 1990s but they tended to last for weeks rather than months – there were no instances of whole winters of unrelenting coldness and blocking. The first week of January 1997 was a good example of how a blocked weather type can maintain consistently cold weather across most of Europe (Box 5.3).

Hazards of heat and drought

The hazards of heat and drought in northern and central Europe tend to be associated with periods of large-scale blocking. Any association between blocking and fine, dry weather is dependent upon the location of the block: if Rossby waves are located such that blocking low pressure is situated over part

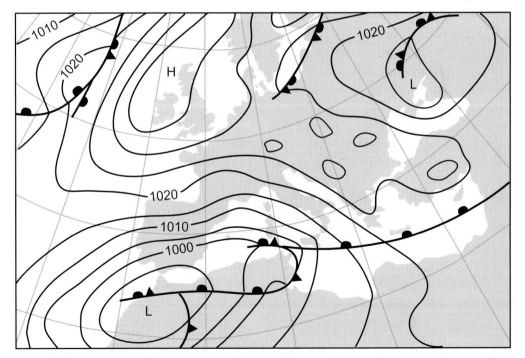

Figure 5.7 *The synoptic weather situation across Europe on 24 January 1947 (18:00 h) showing the extent of the pressure 'reversal' between Scotland and North Africa.*

Box 5.3 Winter blocking in Europe – the first week of January 1997

1997 began with a highly negative NAOI with cold easterly winds covering most of Europe. Pressure ranged from 1037 mbar to the west of Scotland to 1006 mbar over northern Portugal (Figure 5.8(a)). New Year's Day was the coldest day of the month with midday temperatures below −10°C from northern France to Russia. The weather over the Mediterranean was unsettled with two frontal systems separating cold air over central Europe from warmer air to the south: 15°C was recorded in Corsica while Bologna in northern Italy was reporting 0°C with thundery rain.

Over the next week the pressure pattern changed little. Temperatures as low as −18°C and −14°C were reported from Hamburg at 06:00 and 12:00 h, respectively, on the 2nd. There were some sharp weather contrasts within the Mediterranean region. On the 3rd Palermo, Sicily, experienced a southerly föhn wind and a temperature of 21°C at 06:00 h. In contrast, demonstrating the continentality (and altitude) of the central Spanish *Meseta*, Madrid had a midday temperature of just +3°C with snow.

of Europe then wet weather and flooding can result, for example, in central Europe in 2002. However, an upper trough located to the west of Europe will promote the advection of warm southerly winds across the Continent and central Europe may lie under an upper ridge (as in summer 2003).

The origins of dry, hot summers in Europe

Long spells of fine weather in Europe are usually associated with a blocking high pressure system at the surface and a controlling ridge in the upper atmosphere. The upper ridge typically displaces the flow of the polar front jet stream to the north of the average position, for example, between Iceland and Greenland. On occasions, the jet stream may split, sending a southern branch towards the Mediterranean. The key feature is the poleward turning of the mid-latitude westerlies as they approach western Europe.

Warmth can be enhanced by two positive feedbacks. In the summer half-year, once a blocking high has become established over Europe, it is likely to promote warming of the land and the adjacent sea surface. The track of the westerlies around the pole is itself determined by temperature patterns. Areas of unusually warm (cold) air or water can divert the airflow polewards (equatorwards) as it responds to the temperature change. Warmth generated over Europe will shift the latitude of temperature contrast to the northern fringes of Europe and the mid-latitude westerlies often follow this poleward displacement, helping to sustain the warmth over Europe.

The second feedback is the fact that a dry land surface is able to utilize solar radiation as sensible heat – where little moisture evaporates, little heat is lost in the form of latent heat.

Two hot summers separated by a dry winter: 1975–76

The 12 months from May 1975 was the driest such period on record in southern Britain. The key to the persistence of the dry weather lay in the combination of low sea temperatures over the northwest of the Atlantic and warmer water around Europe, sustaining a poleward deviation of depressions. Not since 1749–50 had the period from one summer to the next been so dry. The much-needed

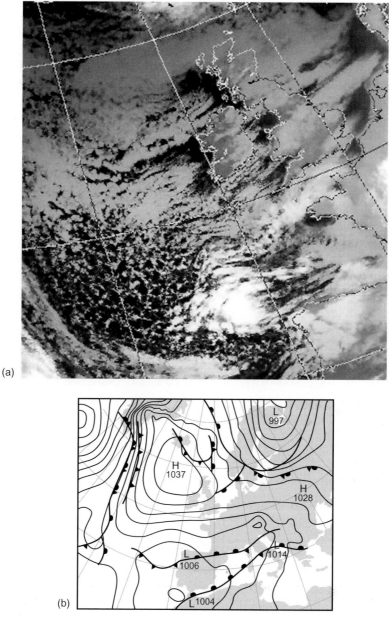

Figure 5.8 *(a) NOAA AVHRR IR satellite image for 08:30h 1 January 1997; (b) synoptic weather map for 12:00h 1 January 1997.*

replenishment of surface and aquifer water resources failed to occur. The record warmth of summer 1976 was associated with a southerly flow developing downwind of a cyclonic area in the central North Atlantic lying just east of an area of relatively cool water.

The fringes of Europe experienced unsettled weather at times as the jet stream split into separate branches south and north of the blocking high. Iceland had an exceptionally wet summer under the influence of the northern branch and there was an unusual frequency of heavy rain and thunderstorms over southern Europe with severe storms in Italy and Spain in August.

A burst of intense heat: summer 2003

August 2003 brought a period of intense heat to much of Europe that followed many months of below average rainfall. After an exceptionally hot June, the land surface was dry and thunderstorms were infrequent. With high pressure continuing through August conditions were in place for record-breaking heat in many countries. Day temperatures often exceeded 40°C in many countries and the warmth of nights, especially in cities, was a major factor accounting for perhaps as many as 14 500 additional deaths in France. In Britain, the national temperature record was broken when 38.5°C was recorded at Faversham (Kent) on 10 August. National temperature records were also established in Germany and Switzerland. By the end of the summer, sea surface temperatures exceeded 30°C in parts of the western Basin of the Mediterranean Sea, a significant store of energy for the following months.

THE CLIMATIC HAZARDS OF THE MEDITERRANEAN BASIN

Climate is an important unifying factor through the Mediterranean region (Perry, 2000b). It is associated with distinctive vegetation types and it provides an important economic resource for the development of tourism. This is a bi-seasonal climate of summer drought and winter rains in which the rainfall of the summer half-year is less than one-third of that of the winter half. Summers are usually described as being warm or hot; winters as mild. However, generalizations and climatic averages should be treated with caution in this climatic zone as it is highly dynamic, with much variability from month to month and from season to season.

The Mediterranean Basin itself is the only part of the world where winter moisture extends a significant distance from the western seaboard of a continent. This is due to the presence of the Mediterranean Sea itself, an important source of moisture and warmth in winter. The Mediterranean climate is found on both the northern and southern shores of the Sea, comprising both southern Europe and North Africa.

Capricious winter weather in the Mediterranean Basin

The two-season climatic pattern hides a diversity of weather types and hazards. The climate may be predictable but the weather often is not. The character of winter weather is dictated by the state of mid-latitude airflow. The description 'mild, wet winters' fails to convey how winters can combine violent, rapidly developing storms and long spells of fine, dry weather.

The winter rains usually start with the first incursion of mid-latitude air into the region. This is brought about by the equatorward drift of the polar front and the mid-latitude westerlies in autumn. The key to the capriciousness is the creation of unstable air; the combination of relatively warm air at the surface with much colder air several kilometres above the surface. The sea surface temperature of the Mediterranean can approach 30°C at the end of summer. The high thermal capacity of water means that heat is retained well into autumn.

As autumn progresses, the likelihood of the air being cooled by a plunge of mid-latitude air increases. This tends to happen abruptly in response to the formation of an upper trough and associated surface low pressure areas (typical synoptic situations are shown in Figures 5.7 and 5.8(a)). An upper trough is an area in which the air above the surface is relatively cool and low pressure centres in the upper circulation are called 'cold pools'. As northwesterly or northerly winds develop in the wake of depressions, the contrast between the warm sea and the colder air above is enhanced because of the faster temperature response of the atmosphere. This breakdown of the summer drought is termed the **autumn break** (Box 5.4). Figure 5.9 shows a typical autumn break storm system over eastern Spain. Even on fine days, rising thermals over Mediterranean islands can lead to the patterns of overland cumulus shown in Figure 5.10.

Box 5.4 Consequences of the 'autumn break'

The 'autumn break' is an erratic phenomenon with little regularity from year to year. The consequences of flash flooding from the intense rainfall are a product of local topography – not only does topography help to channel flood waters towards highly populated valleys but the rugged, mountainous terrain close to the coast of Spain and Italy provides further impetus to lift unstable air. One of the largest cities to be affected is Barcelona (Wheeler, 1988; Tout and Wheeler, 1990). Daily rainfalls exceeding 800 mm were recorded around the Gulf of Valencia in October 1957 and November 1987 and similar amounts have been recorded in the eastern Pyrenees and in Liguria, Italy (Penarrocha et al., 2002). In these regions high ground comes close to the shores of the Mediterranean Sea.

Many other parts of the Mediterranean Basin also experience the autumn break. It is most severe where warm, humid southerly or southeasterly winds blow onto a mountainous coastline from a warm sea surface. In central and northern Italy, some of the worst flash floods also occur in association with summer thunderstorms whilst in the central southern Mediterranean the trigger is often instability caused by North African depressions (see Box 5.5).

The event can be particularly severe when it occurs early in September when high temperatures provide a greater energy source for convective uplift. In 2002, heavy rain and hailstorms caused widespread damage to vineyards in Italy and France. Over 30 people were killed in floods between 8 and 10 September when 650 mm of rain was reported locally in southern France. At the same time, Athens was experiencing daily afternoon convectional storms; over 200 mm fell in Athens in the first 12 days of September compared with an average of just 11 mm for the whole month. In September 2003, Malta experienced severe localized rainstorms when air became destabilized over a very warm sea: sea surface temperatures reached 32°C off Tunisia and on 15 September 226 mm of rain fell at Balzan in little more than 6 hours. These events highlight the intense variability of Mediterranean weather.

Over northern and central Spain and across to northern Italy there is a twin rainfall maximum in spring and autumn; summers here often see heavy convectional rainstorms (this is on the margins of the humid warm summer climate region). However, further northwest, in northern Portugal and along the north coast of Spain, the winters can be very wet. For example, at Vigo, in Galicia on the west coast of Spain, the average rainfall increases from 21 mm in July to 198 mm in January (Uriarte, 1980) and the annual average of 1432 mm is higher than for any city in the UK.

The north coast of Spain has a similarly wet winter climate as Galicia; rainfall totals of as much as 1500 mm on the coast at San Sebastian increase further over the Pyrenees and Cantabrian Mountains. The effect of the mountains on the coastal climate depends on wind direction; if winds blow off-shore then a strong föhn wind may warm the coast, but if onshore winds prevail low cloud may penetrate inland as far as the mountains. The average annual sunshine on the north coast is around 1800 hours (Wheeler, 2001), similar to the English Channel coasts and little more than half the amount of southern Spain.

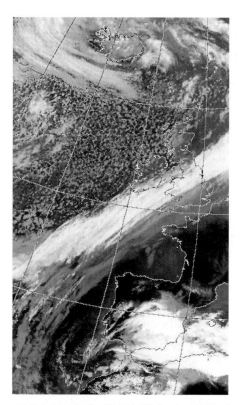

Figure 5.9 *NOAA AVHRR thermal infrared image for 15:11 h 20 October 1982, showing a severe storm system centred over Almeria, southeast Spain, where 425 mm of rain fell in 24 hours. Northwest of the distinct cold front over the British Isles can be seen an array of shower cumulus clouds (the dark grey tone indicates warmth, indicative of shallow cloud in contrast to the brighter tone of the deeper storm cloud over Spain).*

The formation of low pressure centres around the Mediterranean Basin

The Mediterranean Basin has one of the highest frequencies of low pressure systems in the world. Most form within the Basin because:

- the orography of the Basin encourages cyclogenesis by forcing air to ascend over mountain ranges before it can enter the region. The drag on the air exerted by mountains will lead to a loss of momentum. Pressure builds up on the windward side of high ground and decreases on the lee-side. The latter factor encourages the formation of 'lee depressions' – meso-scale depressions. The reduction in air pressure is also encouraged by the fact that it may have become unstable on crossing over the sea;
- surface heating of land encourages air to rise once it has become unstable. This is sufficient to generate thermal lows over inland areas that are especially hot in summer, notably central Spain and (to a lesser extent) Turkey.

There is also an important source of depressions adjacent to the region. North African depressions form in hot air masses over northern Africa, frequently over the Atlas Mountains (they are sometimes called Atlas Mountain depressions). The north coast of Africa is often a sharp thermal discontinuity between warm and cold air, especially in spring, encouraging the formation of depressions. Those forming over the Atlas Mountains tend to move east-northeast; as they pass over the Mediterranean Sea (often near Malta) they acquire moisture and this, combined with the heat energy within the

Figure 5.10 *Towering cumulus rises over Gozo and Malta on a sunny day in April 2000 (all the cloud was due to rapid convection over the centre of each island; adjacent sea areas were clear of cloud).*

systems, can initiate intense rainfalls. These systems usually have only a slight pressure gradient and therefore move slowly, increasing the risk of persistent heavy rain and flash floods (Box 5.5).

Satellite images of Mediterranean depressions reveal that cloud patterns occasionally have a near-circular shape with concentric spirals of cloud. This characteristic, together with their relatively small size (a radius of less than 150 km) has led some to suggest that they resemble tropical cyclones. A recent example occurred in January 1995 when a depression had a distinct 'eye' with peak winds approaching 80 knots (Figure 5.12). On 26 January 1998, a ship lost 37 containers in Force 11 gales in the vicinity of one such storm (Burroughs and Lynagh, 1999). It appears that these systems are triggered by parent depressions in unstable air and are sustained by the release of latent heat through the condensation of water vapour (Pytharoulis *et al.*, 2000).

Campins *et al.* (2000) have confirmed that the frequency of depressions in the region reaches a peak in summer (29% of the annual total). While this may seem to be at odds with the dryness of the Mediterranean summer, the heating of air in summer leads to a westward retreat of the Azores High and the development of thermal lows over land. The air is usually too dry to initiate condensation of water vapour even when pressure is low. The minimum frequency of depressions is in winter when air pressure reaches a peak.

Box 5.5 North African depressions as climatic hazards: storms, landslips and flash floods

The waters between Malta and Africa can provide massive amounts of energy to passing weather systems, particularly North African depressions. The North African flood disaster of September 1969 resulted in the deaths of 300 people in Tunisia and Algeria. This was caused by a North African depression that moved slowly east towards Malta; as it acquired additional energy from the warm sea it turned back towards North Africa as an invigorated system.

Over half of the heaviest daily rainstorms to affect Malta are caused by North African depressions. The change from summer drought to intense rain can be an important agent of geomorphological change. Short-lived storms cause large amounts of runoff; 103 mm of rain fell between 22:20 and 24:00 h on 22 September 1997, one of several storms that resulted in landslips along spring lines where impermeable strata reach the surface (Figure 5.11).

Figure 5.11 *The site of a landslip on Gozo in the Maltese Islands caused by the heavy autumn storms of 1997. The slip occurred where a spring line on blue clay intersects a road.*

Winter drought in the Mediterranean

In most Mediterranean countries a large proportion of the total annual rainfall tends to fall on a small number of wet days, highlighting the importance of convection as a dominant rainfall-generating mechanism. Uplift of unstable air can be fast, dramatic and unpredictable, occurring just when the air is sufficiently unstable.

Air pressure in winter is typically higher than in summer. Most winters have several spells of settled, mostly dry weather in which the Azores High extends right across the region. These spells generally coincide with a progressive airflow in the mid-latitudes when the Azores High is at its most intense

Figure 5.12 *Satellite image of a depression close to Malta on 16 January 1995 that had some features of a tropical cyclone (a central 'eye' and an absence of frontal cloudbands).*

and is anchored to southern Europe. In Greece, these periods of fine winter weather are referred to as Halcyon Days, when the mythical halcyon bird built its nest and hatched its young (Perry, 2000b).

Since the early 1980s there have been several winters in which progressive airflows have persisted over Europe, maintaining dry weather in the Mediterranean and raising concerns about water use and availability. Water supply problems can be particularly acute in agriculture and soil moisture is the dominant meteorological control on yields of many crops. The mildness of the Mediterranean coastlands has been exploited for the production of early vegetables and several harvests of many crops are possible each year if irrigation water is reliable.

5.3 GEOGRAPHICAL PATTERNS OF WEATHER IN THE BRITISH ISLES

WEATHER, CLIMATE AND TOPOGRAPHY
The British Isles experiences a wide variety of weather types because:

- air masses can be either continental or maritime in origin;
- the distribution of high ground produces a variety of patterns of exposure and shelter as wind directions vary;
- all airstreams pass over sea surfaces and are modified to differing extents.

The aim of this section is to examine the range of weather contrasts that accompany changes in surface airstream direction, with illustration by satellite imagery. Two climatic influences become stronger towards the northwest: proximity to the average route of North Atlantic depressions and the maritime influence of the North Atlantic. These influences are enhanced by geographical factors such as the concentration of high ground in the northwest. Taking the year as a whole, these climatic and geographical factors promote a southward gradient of increasing temperature and sunshine and an eastward gradient of dryness (Mayes and Wheeler, 1997).

WEATHER TYPES AND REGIONAL CLIMATES

The regional climate and weather associated with particular airstream directions arises both from the original characteristics of each airstream (air mass) and from modifications caused by the underlying surface. Original characteristics are determined by whether the source of the air is over land or sea. The route taken by different airstreams then alters the thermal structure of the air, modifying instability. It also determines the way in which air interacts with topographic features creating local weather variations.

Southwesterlies and westerlies: 'west is wet'

This is the prevailing wind direction, occurring usually on over 100 days per year. 'West is wet' is true in two senses. First, westerly winds are moisture laden as a result of their source over a large expanse of ocean. This is the direction from which the most vigorous depressions travel, generating large amounts of frontal rain as this moist air is forced to rise in the circulation of these weather systems. Second, the proximity of high ground to many western coasts yields large amounts of orographic rain as moist air is abruptly lifted (Chapter 2). Orographic rain often combines raindrops with drizzle, providing a dense mist of precipitation that has a high capacity to drench those outside. Whilst reinforcing the intensity of frontal rainfall, it also has a tendency to continue through the warm sector of depressions, between the warm and cold fronts, giving highly persistent heavy rain and drizzle. The wettest weather in the northwestern uplands occurs when heavy rain becomes persistent in wide warm sectors. Figure 5.13 shows an example of this situation when the hills around Loch Lomond, north of Glasgow, recorded 167 mm of rain in 24 hours from 09:00 h on 10 December 1994 and serious flooding followed (Black and Bennett, 1995).

The western uplands are wetter than rain-shadow locations by a wider margin than is often realized. For example, in January 1989, monthly rainfall totals ranged from 855 mm in Glenshiel in the Western Highlands to just 4 mm at Fyvie Castle in Aberdeenshire. Daily rainfall totals over the 48 hours ending 09:00 h on 7 February ranged from over 250 mm on the Western Highlands (306.1 mm at Kinloch Hourn, a Scottish record for 48 hours) whilst the east coast from Aberdeenshire south to the Borders had less than 5 mm (Roy, 1997).

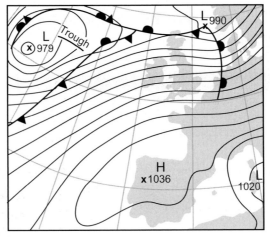

Figure 5.13 *Synoptic weather map for 10 December 1994 showing persistent, moist warm sector airflow over Scotland.*

Air that originates from tropical maritime air masses is relatively warm and thus becomes stable as it cools on approaching the British Isles. It takes little uplift to saturate these usually moist onshore winds; southwest facing coasts and hills are prone to advection fog and drizzle. However, after this moisture has been lost following ascent, the air descends into the rain-shadow areas, sometimes promoting the föhn effect. Temperature peaks can correspond closely to areas of bright weather in the lee of high ground, as shown by an exceptionally mild spell in February 1998 when 19°C was recorded at Minehead (to the lee of Exmoor) on the 13th and 14th.

If air is of polar origin (polar maritime air mass) the patterns of cloud and weather will resemble those of northwesterlies (see below). This weather type will be more unstable and bring showers instead of sheets of cloud.

Northwesterlies and northerlies – unstable air brings showers

These winds are usually of polar maritime origin and tend to bring air that is unstable by the time it reaches the British Isles. This is because the lowest air (which is more sensitive to surface temperature) becomes warmed more than the air several kilometres up, increasing the lapse rate. Unstable air creates thermals, cumulus cloud and showers, associated with the characteristic bright, clear weather that follows the clearance of a cold front.

The distribution of showers reveals much about the pattern of heat at the surface as shower cumulus forms over the warmest surfaces. In the winter half-year (and more generally at night), showers form most readily over the relatively warm sea around the British Isles. Coasts exposed to the northwest or north may have frequent showers (Figure 5.14). As the winds pass

Figure 5.14 *Instability over the sea. Cumulus distribution typical of northwesterly airstreams of the winter half-year or at night; NOAA AVHRR thermal infrared image for 06:29 h, 12 April 1998. The cumulus cloud dissipates over Ireland and England in response to the lower surface temperature of the land compared with the relatively warm sea surface.*

inland, over the cooler land, showers gradually die out, leaving clear skies. So, while Liverpool may have many heavy showers, places such as Dundee, Cardiff and Torquay may have bright sunshine and remain dry.

In summer, the land is warmer than the sea during the day and consequently the heaviest showers will develop. These may culminate in thunderstorms where there is an additional stimulus to uplift, such as an urban area or a sea-breeze front (see Box 5.6).

Northerly winds often bring similar weather to northwesterlies. The geographical distribution of the showers may differ in that the North Sea coasts of Scotland and England may now have onshore winds, increasing the risk of showers here. Northerly airstreams are particularly associated with a potentially hazardous meso-scale weather feature; the polar low. These appear on satellite images as clusters of deeper cumulus cloud. They are a feature of cold polar airstreams and tend to be associated with sudden heavy snowfall, especially over northern hills. They have been relatively infrequent since the 1970s, though several gave snow to many parts of Britain in April 1999 (Bowker, 2000).

Scotland suffered a particularly cold end to 1995 as a northerly airstream became established on the eastern flank of a high over Iceland. Snow showers from around 19 December provided a cover of snow, especially in the north and east. On 29 December, Glasgow airport recorded its coldest day on record ($-12°C$) and the following night the temperature at Altnaharra automatic weather station in the northern Highlands dropped to $-27.0°C$ (Rowley-Gillard, 1996). The standard minimum thermometer read $-27.2°C$ (Burt, 1997), equalling the lowest air temperature recorded in the UK (at Braemar, on 11 February 1895 and 10 January 1982). As milder air spread in from the south later that day, the temperature rose above freezing point, resulting in the largest diurnal range in temperature yet recorded in the British Isles (29.3 deg. C).

Winds from a northerly quarter do not always bring unstable air. If air has circulated around the northern side of an anticyclone, the origins may not be from the north at all. Instead of bringing cumulus cloud and showers, wide areas of Britain may be covered by stratocumulus. The subsidence

---

Box 5.6 The 'peninsula effect', sea-breeze fronts and the 'estuary effect'

Satellite imagery has enabled weather forecasters to examine the patterns of showers and cumulus clouds in detail. The 'peninsula effect' is the horizontal convergence of air towards the centre of peninsulas, i.e., away from the coasts. The landward penetration of sea-breezes is usually marked by the sea-breeze front; a line of cumulus cloud parallel to the coast. Sea-breeze fronts from opposing coastlines can meet over the centre of peninsulas, causing an enhanced band of cumulus. This can be seen in Figure 5.15 as a line of cumulus over southwest England. Coastal fringes, now in the cool air brought by the sea-breeze, become clear of cumulus; this can be seen along the coast of west Wales and over northern France. Heavy showers formed over southeast England – the image shows the large tops of cumulonimbus cloud drifting over the English Channel (daytime showers get 'exported' from land to sea where the wind is off-shore).

The 'estuary effect' is the winter-time counterpart of the 'peninsula effect' when, in autumn and winter, estuaries are warmer than land, and often warmer than the open sea. Convection may thus become enhanced and, as with the peninsula effect, the resulting showers may be boosted further by horizontal convergence of air. Autumnal showers entering the Bristol Channel on west-southwesterly winds may be particularly heavy around Cardiff, Bristol and north Somerset. The Thames estuary produced a similar effect (in northeasterly winds) between 11 and 14 January 1987 when snow showers intensified as they passed over the water; the area around Maidstone had up to 52 cm of snow, the heaviest for 40 years.

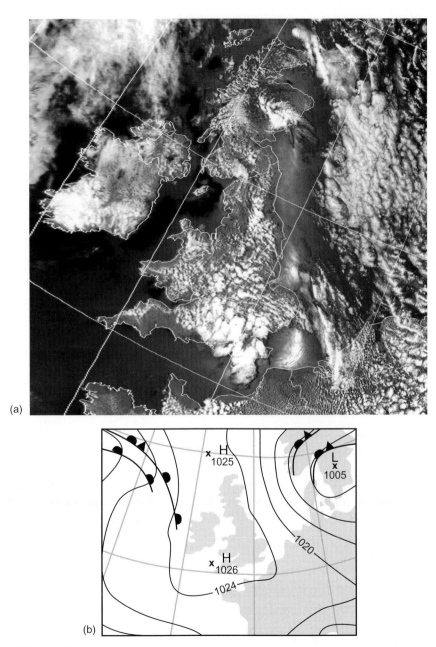

(a)

(b)

Figure 5.15 *Instability over land. Build-up of cumulus over land and the effect of sea-breeze fronts: (a) NOAA AVHRR visible image for 14:25 h 26 June 2000; (b) surface synoptic chart for 12:00 h.*

of air around an anticyclone tends to suppress the uplift of cumulus – stratocumulus is often formed from the spreading out of cumulus cloud when air becomes less stable. Places downwind of high ground such as Cardiff and Plymouth may be the last to see such cloud in northerly winds and remain relatively bright and sunny. Further north the weather may be overcast, cool and gloomy.

Easterlies and northeasterlies: stable air brings 'haar'

These winds are associated with blocked weather patterns and high pressure systems often located close to Iceland and the Norwegian Sea and low pressure over central Europe. It might be thought that northeasterly winds would bring unstable air because of their usually cold source. The reality is that the North Sea usually generates a cooling effect on the lowest layers of air and this ensures that these airstreams are usually stable by the time they reach the east coast. If the air is cooled to dewpoint temperature, condensation will produce stratus or advection fog; the 'haar' of northeast England (Wheeler, 1997). This forms over the sea and drifts with the wind onto east-facing coasts where it may later evaporate if land surface temperature exceeds dewpoint temperature (Figure 5.16).

The distribution of cloud is sensitive to both surface temperature and any blocking by high ground; Figure 5.17(a) shows that it can spread as far west as the eastern side of the Welsh mountains following a cool night, even in July. The daytime heating of the land causes the leading edge to evaporate and it is usually much less extensive over land by afternoon (Figure 5.17(b)). While no sunshine was recorded over Yorkshire, Manchester (sheltered by the Pennines) had 11 hours of sunshine and the highest temperature of the day was at Keswick in the Lake District at 25.3°C. Tiree in the Outer Hebrides was the sunniest location with 13.1 hours, characteristically favoured in easterly airstreams. On the following day Keswick reached 30.3°C and the sunniest place was again in the Outer Hebrides: Stornoway, with 15.3 hours.

A more unusual effect of shelter in a broad area of stratus was noted by Pedgley (1999) over the Shetland Isles. On 19 July 1997 a distinct clear zone downwind of Shetland was noted, caused by a combination of airflow blocking and warming over the land.

Southerlies and southeasterlies – blow hot or cold

The weather of the British Isles acquires continental characteristics in southeasterly airstreams when high pressure is typically located over Scandinavia with low pressure close to the Bay of Biscay. This weather type characterizes our warmest summers and coldest winters. Lowest day temperatures in winter are often found in the far southeast of England where the airstream has the shortest track over the warmer sea (e.g., in January 1997 – Box 5.3). This short sea fetch also means that the risk of low cloud or haar is less than for easterlies and is more likely on the east coast of Scotland rather than England.

A change in wind direction to southerly changes regional weather types. The moisture content of air reaching the west of the British Isles increases because the air now has a maritime source and south-facing coasts may become overcast and misty. This is because southerlies are usually stable airstreams and mist, fog and drizzle may form when the surface layers are cooled to dewpoint temperature on their northward travel (Figure 5.18). By contrast, southerlies may bring hot, unstable air northwards – the 'Spanish plume' of summer heatwaves (Box 5.7).

Southerlies in Cork, Plymouth and Swansea are not necessarily providers of great warmth or sunshine. Indeed, southerly weather here can be very wet. By contrast, as with other wind directions, places sheltered by high ground are favoured with more sunshine and higher daytime temperatures. November 1994 was dominated by moist southerlies. It was the dullest November on record in the south of Ireland but the sunniest on record on the north coast (Mayes and Wheeler, 1997).

Figure 5.16 North Sea 'haar' on 17 March 2003 (NOAA AVHRR visible image for 14:13 h). Stratus formed widely over the cold surface of the North Sea (typical of spring), evaporating after passing onshore along the east coast of England and Scotland.

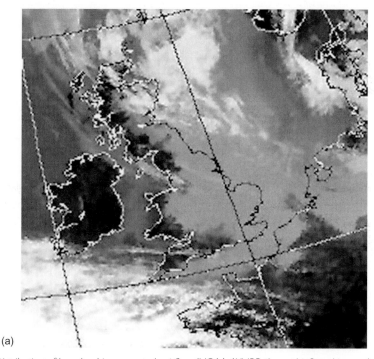

(a)

Figure 5.17 Distribution of low cloud in an easterly airflow (NOAA AVHRR thermal infrared images) (a) 08:21 h. Distribution of low cloud in a northeasterly airflow (NOAA AVHRR thermal infrared images).

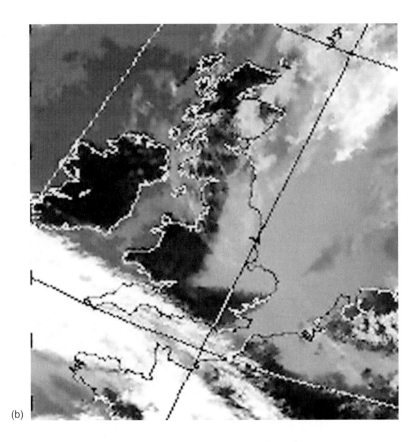

(b)

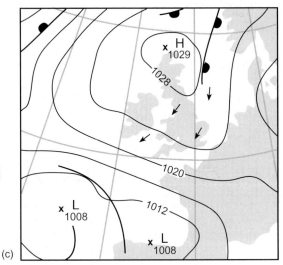

(c)

Figure 5.17 (Continued) (b) 15:51 h. (c) Surface synoptic chart for 12:00 h on 22 July 2000. By afternoon the stratus extended only as far west as the Pennines and the east Midlands. Note how it is possible to distinguish low level stratus (grey) from the warmer land (black). The distribution over the cooler sea was less clear, though it was extensive over the North Sea on this occasion. The higher, deeper frontal cloud approaching southern England appears lighter as a result of lower cloud-top temperatures.

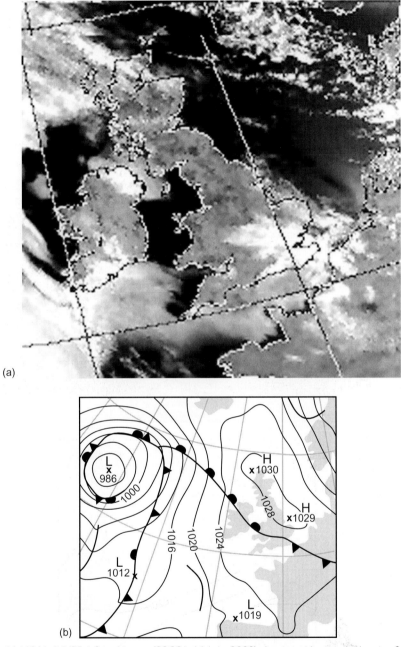

(a)

(b)

Figure 5.18 *(a) NOAA AVHRR infrared image (08:29 h 16 June 2000) showing widespread advection fog in the English Channel and to the south of Ireland. (b) Surface synoptic chart for 12:00 h.*

Southerlies frequently lead to the föhn effect developing to the lee of high ground. On 12 March 1957 the temperature at Cape Wrath – the furthest northwestward extremity of the British mainland – reached 23°C, a value equivalent to the highest temperature reached there in a typical summer. The föhn effect is particularly associated with the Moray Firth in Scotland and the coast of North Wales. The highest January temperature recorded in the UK (18.3°C) was recorded at Aber (downwind of Snowdonia) on 10 January 1971 (Pedgley, 2000).

Box 5.7 Meso-scale convective systems and 'Spanish plumes'

As showers develop into thunderstorms, the area of cloud visible from space increases as the cirrus outflow plume at the top of a cumulonimbus cloud spreads out. Once separate, cloud systems can merge, creating meso-scale cloud systems with much more extensive cloud plumes. Identification on imagery is important because these systems can develop into major thunderstorms and satellite tracking is an important tool in short-term prediction (Nowcasting) of the storms.

A meso-scale convective system is defined as an area of precipitation of more than 100 km extent horizontally (Gray and Marshall, 1998). While being the major provider of tropical rainfall, they also develop over Europe, especially when warm southerly airstreams become highly unstable air. They tend to be associated over the British Isles with southerly airstreams in summer, which are destabilized by colder air to the west lying behind an eastward-moving cold front. These southerlies often run from Spain to the British Isles, initiated by strong heating on the Spanish Plateau – the 'Spanish plume' (Figure 5.19). The contrast of wind direction enhances the temperature contrast either side of the front (warm advection in the southerlies, cold advection in the westerlies). The event of 24 June 1994 was of sufficient size to be classed as a meso-scale convective complex and produced severe thunderstorms across a wide area (Young, 1995).

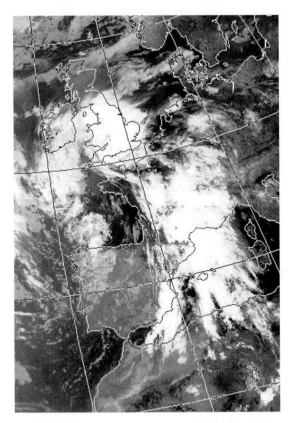

Figure 5.19 *The Spanish plume revealed: NOAA AVHRR infrared image for 03:39 h 1 June 1983. Cloud-top temperatures of the cumulonimbus cloud shield reached around −60°C; the clusters have a distinctive smooth oval shape.*

Anticyclonic weather – fine with occasional gloom

The fine weather associated with anticyclones is caused by subsiding motion of air. Though the evaporation of cloud is caused by the warming of air on its descent, it is misleading to associate 'highs' with high temperatures. High pressure areas encourage temperature extremes. Low cloud amounts promote heat loss at night and warming by day. Warming is encouraged further by the drying of the land surface in anticyclonic spells; wet surfaces are slow to warm because energy is required to evaporate surface moisture before it can be used to heat the surroundings (latent heat of vaporization is absorbed from the local environment).

The temperature extremes that can develop under anticyclones are ameliorated by the high thermal capacity of water, which makes coastal districts slow to cool in autumn and slow to warm in spring and summer. The low mixing rates of air in light winds allows sharp local temperature contrasts to develop. For example, on 30 April 1993 the maximum temperature beside the North Sea at Tynemouth (Tyne and Wear) was only 8°C while a few miles inland at Newcastle the temperature reached 17°C; further inland still, Penrith (west of the Pennines) reached 25°C.

The failure of cumulus clouds to develop on occasions (see for example Figure 1.7) can be explained by the presence of stable air at or close to the surface in association with a temperature inversion, acting as a cap on thermals, and low relative humidity. In a subsidence inversion, stratus or stratocumulus may form in the cool layer of air close to the surface while higher layers will lie in the warmer subsiding air. The inversion level may drop at times to lie below mountain summits and it is on these occasions when the mountain peaks of Britain experience some of their most exhilarating weather – bright sunshine and brilliant visibility may occur above the inversion level in contrast to the gloom below.

Cyclonic weather – stormy or blocked

Cyclonic weather is perceived as stormy, unsettled and wet; in reality, conditions depend on the energy of the weather system. High energy – characteristic of progressive weather – is marked by a rapid passage of the 'low', stronger winds and abrupt weather changes as fronts cross the country; the wet winter of 2000–2001 was a good example (Marsh, 2001).

All low pressure systems eventually lose energy and dissipate. As thermal contrasts weaken, uplift at both the depression centre and along fronts weakens. Air pressure will increase and winds thus become lighter. In blocked weather patterns, low pressure may stagnate over the British Isles and fronts often become slow moving and weak. A stationary active front may give large amounts of rain locally but, more typically, the weather is quiet with light winds and slow-moving showers. In summer these may be heavy and thundery over land and severe thunderstorms usually occur in cyclonic rather than anticyclonic situations.

5.4   THE CLIMATES OF NORTH AMERICA

North America presents a very different obstacle to mid-latitude westerlies compared with Europe and this difference is expressed in a different pattern of weather and climate. The major topographic feature – the Western Cordillera (Rocky Mountains) – runs north–south rather than east–west. This means that the maritime influences of the Pacific coast are blocked after quite a short passage over land. The western side of the Rockies defines the boundary between a maritime west coast climate to the west and a variety of continental climates to the east.

WEST COAST MARITIME CLIMATES

The Pacific coast of Canada and the USA lies in a similar location in relation to the mid-latitude westerlies as western Europe and the western Mediterranean Basin. As latitude decreases, the length and intensity of the winter rainfall diminishes and summer drought lengthens, culminating in a Mediterranean climate type in much of California (Figure 5.20). Indeed, the contrast between wet winters and dry summers is generally greater than in western Europe, a difference attributable to the relatively cool ocean currents off-shore. The region does not experience the advection of moist, warm air that can happen in Europe and further east in North America. The most distinctive feature of the coastal climate is the persistence of advection fog along the coast. In the absence of extensive water bodies, areas inland become increasingly arid, despite an increase in elevation.

CONTINENTAL CLIMATES OF NORTH AMERICA

Continental climates cover most of North America east of the Western Cordillera.

Geographical controls on weather – upper airflow

The Western Cordilleras (Rocky Mountains) extend from Alaska to Mexico, from 60°N to 30°N. This is a sufficiently large barrier to disturb the characteristics of the mid-latitude upper westerlies – a barrier so large cannot be avoided. Air is forced to flow over the peaks and is squeezed (horizontally), forcing the air to accelerate. This disturbs the direction of the upper winds because the balance controlling their direction is disturbed; the upper winds (the geostrophic wind) result from the poleward pressure gradient force and the equatorward Coriolis force (Chapter 2). The latter is proportional to air velocity and is thus forced to strengthen. The upper westerlies therefore turn equatorward and now run southeast instead of east, forming the Eastern Canadian Trough (Figure 5.21).

As air comes out of the trough it flows northeastwards over the eastern seaboard; this is because of a weakening of the Coriolis force at lower latitudes, a fact that keeps mid-latitude air over the mid-latitudes! The southwest to northwest direction of the airflow (and the jet stream) over this area is nearly parallel with the east coast; this alignment may place tropical air over the Atlantic and polar air over the land. This enhances winter temperature contrasts with serious implications for the winter climate of the East Coast region.

Geographical controls on weather – surface airflow

The influence of surface westerlies is reduced by the blocking effect of the Rockies and by the creation of northwesterlies in the upper atmosphere. The main geographical factor lies in the absence of blocking high ground running east–west giving free rein to northerly or southerly winds. As a result, many meteorological battles occur between polar and tropical air masses as they try to conquer territory over the flat terrain of the Great Plains and the Mississippi valley.

The battleground at any one time is shown by the location of the polar jet stream (Figure 5.21). However, the weather on any day may depart markedly from the climatic average where such contrasting forces battle for territory. A typical source region for polar air may be northern Canada. Tropical air, by contrast, tends to have an oceanic origin in the Gulf of Mexico or on occasions the Atlantic coast off Florida or the Carolinas. Tropical air masses are therefore an important source of moisture, in contrast to the dryness of westerly winds after their descent from the Rockies.

In January, the mean temperature is below −35°C in the far north of Canada, around −10°C in much of southern Canada, more than 10°C along the Gulf coast and as much as 20°C around Miami. In July the extreme values range from less than 10°C on the north coast of Canada to more than 30°C in southwest Texas. Places in between can be subjected to great extremes of temperature from week to week and even from day to day, as will be illustrated in the following regional overviews. It is the resulting contrasts that provide the energy for thunderstorms, tornadoes and other storms that characterize the energetic climatic hazards of much of continental North America.

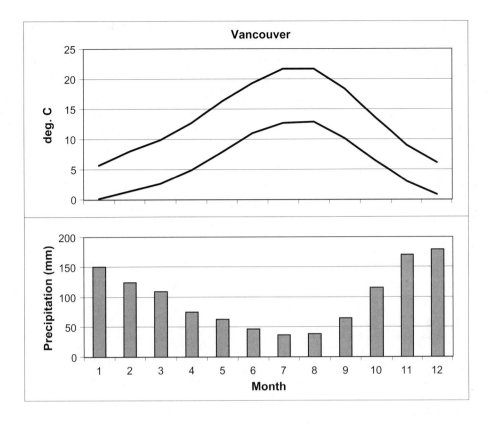

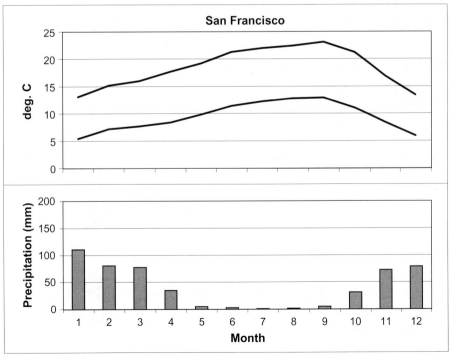

Figure 5.20 *The transition from mid-latitude west coast to Mediterranean climates; Vancouver and San Francisco climate graphs.*

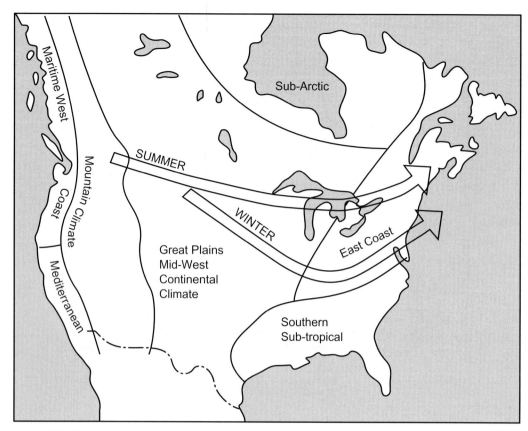

Figure 5.21 *Climatic regions of North America.*

CLIMATIC REGIONS, HAZARDS AND EXTREMES

The Great Plains/Mid-West

As a large area, extending from northwest Texas to the Canadian states of Alberta and Saskatchewan, it might be expected that climates would vary because of the wide range of latitudes. However, this region has two unifying features; it lies to the lee of the Rocky Mountains and occupies the continental interior of North America.

Monthly means (Figure 5.22) reveal little of the variability of temperature that occurs in this area. This zone is one of the great battlegrounds of contrasting air masses. Cold waves occur when a temperature drop of at least 11 deg. C occurs within 24 hours. The 'Alberta Clipper' is the name given to the flow of polar air out of Alberta. Conversely, the föhn winds known locally as Chinooks have been known to raise temperature by as much as 22 deg. C in 5 minutes.

Summer heat-waves are a predictable yet deadly climatic hazard of much of North America. In 1995, over 1000 people died across the USA, over 700 of these in Chicago alone. Each of the seven worst affected cities are in the Northeast or Mid-West where, typically, an extra 1100 heat-related mortality cases occur in an average summer.

Rainfall decreases steadily from southeast to the west; it averages >1200 mm in the Lower Mississippi Valley but less than 250 mm at the foot of the Rockies in Colorado (e.g., Denver in Figure 5.22). This decrease towards the rising ground of the Rocky Mountains highlights not just the strength

148

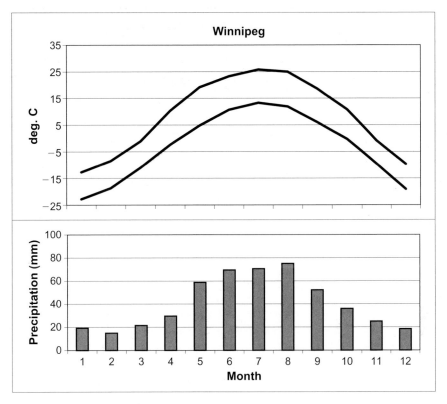

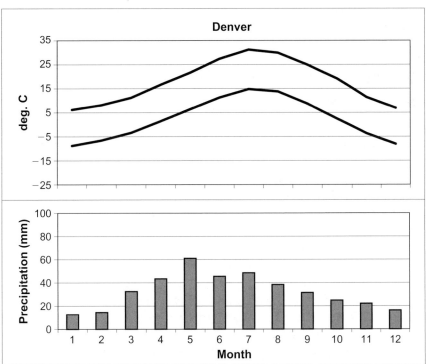

Figure 5.22 The climate of the Great Plains: climate graphs for Winnipeg and Denver.

Box 5.8 The thunderstorm hazard in the Great Plains

Thunderstorms tend to be located close to the edge of the tropical air mass, along squall lines, lying typically just southeast of cold fronts (which are frequently aligned southwest to northeast as shown in Figure 5.23).

The most severe thunderstorms develop when warm, moist and unstable air in the lowest kilometres of the troposphere comes into contact with cold dry air above (e.g., at about 5–6 km). The latter airstream often diverges horizontally and this tilts to the top of the storm clouds so that the falling rain does not fall through the mass of cloud, suppressing the rising thermals by cooling. This divergence therefore increases the duration of thunderstorms, increasing the risk of hail. The pattern of updraughts and downdraughts around each storm system eventually reaches a state of balanced equilibrium. As a result, supercell storms can form, often forming part of meso-scale convective complexes. Severe downdraughts can trigger their own hazard – severe gusty winds called downbursts.

of the rain-shadow but also the importance of the Gulf of Mexico as a source of moisture and rainfall for much of the central and eastern USA. Spring is the wettest season for the moister parts of this region, towards the Lower Mississippi Valley. Warm, moist air from the Gulf of Mexico is now able to move further north than is possible in winter. This air becomes increasingly unstable as it moves closer to colder polar air. This instability can be explosively released in certain weather situations leading to the formation of severe thunderstorms and tornadoes.

Weather extremes in an individual season are often dictated by the location and strength of the polar jet stream. If it remains in a fixed location for several weeks, these areas may accumulate high rainfall amounts from the succession of depressions or associated slow-moving fronts. A notable case was in the Mississippi flood emergency of the summer of 1993. Rainfall in June and July exceeded twice the average in the middle Mississippi Valley, the southwest of the Great Lakes and the central and northern Great Plains. A strong and near-stationary jet stream acted as a 'moisture pump' across these areas (Lott, 1994). Meso-scale convective complexes provided intense falls of rain at times. April to July rainfall exceeded average annual rainfall in places. The economic cost reached US$21 billion, 48 lives were lost and record river flows were recorded. However, the concentration of the jet stream over this central area of North America reduced the frequency of unsettled weather elsewhere. The southeastern USA experienced a dry, hot summer with less than half of average rainfall in June and July in places where the summer as a whole was the driest on record.

The tornado hazard

Tornadoes are the most violent of atmospheric hazards and the most serious in terms of damage in the USA. The focal point for development is often similar to that of the worst thunderstorms; a cold front aligned southwest to northeast, separating warm moist air from cool, dry air. The boundary is called the **dry-line** and this feature, identifiable on satellite imagery (see Figure 5.23), can be a focus for turbulence. Hot air rising from the surface can break through this dry-line, resulting in explosive supercell thunderstorms. Updraughts can reach 100 mph.

As this rising air reaches the cool, dry upper westerly winds, a spinning motion may develop because of wind shear associated with contrasting wind directions. This rotation starts as an extensive circulation covering several kilometres – a meso-cyclone. Continued supply of heat and moisture

Figure 5.23 Three major thunderstorm cells aligned along 'Tornado Alley' on 9 October 2001. The image is from GOES-12 (courtesy of NASA). 23 tornado sightings were made on this date.

Box 5.9 The Oklahoma tornado of May 1999

76 tornadoes were reported from the southern Great Plains states on 3 May 1999. The largest of these (F-5 on the Fujita–Pearson tornado scale) remained on the ground for 4 hours, running northeast through Oklahoma City, resulting in a damage swathe half a mile wide. 54 lives were lost and over 8000 properties were damaged or destroyed in Oklahoma state. Total losses were expected to reach US$1 billion. This illustrates all too clearly how human and economic losses created by tornadoes are as much a function of their track as their magnitude.

---

causes the vortex to narrow. The rotating column of air then starts to lengthen, emerging from the 'wall-cloud' of the meso-cyclone. A funnel cloud is formed as the rotation initiates condensation – it is this condensation that makes funnel clouds visible. The rotating column lengthens downwards; if it reaches the ground it becomes known as a tornado.

The highest frequency of tornadoes in North America occurs in 'Tornado Alley', a broad corridor of vigorous activity, extending through Texas, Arkansas, Nebraska, Iowa, Mississippi, Alabama and Oklahoma (Figure 5.23). The latitude of tornadoes is guided by the seasonal meanderings of the jet stream; they occur in the southern states in winter and as far north as the Great Lakes in summer. In Texas alone there were 475 recorded deaths from tornadoes between 1950 and 1994. Over the USA as a whole, there was an average of 230 deaths per year between 1916 and 1953, with the majority of deaths being associated with the severest tornadoes (Oliver and Hidore, 2002). The trigger of instability is demonstrated by peak occurrences between 14:00 h and 20:00 h and in April and May.

Tornadoes' strength is very difficult to measure accurately, partly because the peak winds are so localized and because anemometers become damaged. However, the Fujita–Pearson scale ranks tornadoes from 1 to 5, where the latter category corresponds to winds of between 227 and 276 knots.

As it is very difficult to forecast the damage track of an individual tornado, it is more realistic to attempt to forecast the areas where multiple tornadoes may develop after the first ones have 'landed'. 148 tornadoes were observed in a multiple outbreak on 3 April 1974 (Oliver and Hidore, 2002) but more may have formed in remote areas and may not have been recorded (Figure 5.23).

Radar assists tornado forecasting and warning services. Conventional rainfall radar can identify hook shapes to rainfall areas; this is often a sign of rotation in a meso-cyclone. Doppler radar can identify whether an object is approaching or moving away by detecting changes in signal frequency. This can identify rotation just before tornadoes form and can then assist in tracking of the resulting system.

The South

The climate of the southern states of the USA (from the coast of Texas to the Carolinas) is transitional between that of the Great Plains and the tropics. Indeed, southern Florida has a tropical climate. The weather is warm and humid for much of the year and proximity to warm coastal waters is clearly an important feature of the climate. The polar air masses rarely reach this far south but short-lived winter cold snaps do occasionally occur and cause serious damage to agriculture, especially to the citrus crops of Florida. Mean temperatures in July widely exceed 25°C.

While the Gulf States east of New Orleans average more than 1600 mm per annum, the westward increase in aridity ensures that average rainfall in west Texas diminishes to under 200 mm. This is despite the proximity to the Gulf of Mexico, normally an important moisture source for the south. Summer rainfall decreases with increasing distance from the Gulf of Mexico. This is partly due to the effects of lower temperature and humidity but also to the influence of high pressure in the upper troposphere. These inland states have a winter–spring rainfall maximum coinciding with the southernmost travel of the polar jet stream.

The dominant climatic hazard of this region is the tropical cyclone or hurricane (see Section 6.3). It is unfortunate for North America that this is one region where tropical cyclones extend their influence out of the tropics towards the mid-latitudes. Tropical cyclones will tend to track northwards away from the tropics if low pressure is present. This is possible over warm land areas even at sub-tropical latitudes. A clockwise-curving track may take these systems along the eastern seaboard of the USA. In September 1999, the Carolinas were affected by three severe tropical cyclones within three weeks (see Section 6.3). The warm, moist air can be an important energy source for mid-latitude depressions, which can go on to give downpours in the northeastern states and Canada (e.g., Hurricane Isabel in September 2003; Figure 5.24) and may continue to develop as they cross the Atlantic (e.g., the origins of the Great Storm of 1987 over Great Britain).

Many of the summer droughts and heat-waves over North America in the 1990s have been in the southeastern states (Table 5.2). Most of the deaths have been attributed to heat-related health stress while the economic losses have been primarily due to agricultural losses.

The east of Canada and the USA

The eastern seaboard of North America has a climate in which temperature is controlled by the land-mass to the west but rainfall is controlled by the influence of the North Atlantic. The influence of the land on temperature is attributable to the prevailing mid-latitude westerly winds, resulting in a far more continental climate than is found on the west coast of North America.

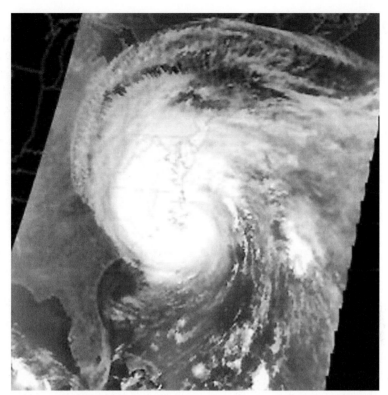

Figure 5.24 *A NASA Terra composite simulated visible image of Hurricane Isabel (11:50h EDT on 18 September 2003). This was the first Category 5 Atlantic tropical cyclone on the Saffir–Simpson scale (Section 6.3) since 1999. This image (courtesy of NASA) shows the unusually large diameter of the system as it started to evolve into an extra-tropical depression after landfall on the eastern seaboard of North America. A State of Emergency was declared in Virginia and Washington DC because of flooding.*

Table 5.2 Location and impact of major droughts and heat-waves in the USA, 1986–2001

Year	Region	Deaths	Economic losses (1998 prices) US$ billion
1986	Southeastern states	100	1.6–2.4
1988	Central and eastern states	5000–10000	56
1993	Southeast states	>16	1.0
1998	Texas to Carolinas	>200	6–9
1999	Eastern states	502	>1.0
2000	South central and southeast	140 (nationwide)	>4.0

Source: US National Climate Data Centre

The monthly temperature graph for Boston (Figure 5.26) shows that proximity to the Atlantic has only a small ameliorating effect on temperature range compared with locations further inland such as Toronto. The fact that the temperature at night does not, on average drop below 20°C at Washington and New York at the height of summer contributes to the discomfort of the summer heat and humidity, increasing the demands upon air conditioning and the medical services. Adverse health effects of the climate are sometimes aggravated by large day-to-day temperature variations.

Box 5.10 The Great Lakes snows – a vivid illustration of atmospheric instability

The 'lake effect' snows of the Great Lakes are one of the most vivid examples of the local effects of unstable air. Convective precipitation is triggered by vertical temperature contrasts rather than by heat itself. Since large lakes will be warmer than the surrounding land areas in winter (owing to the high thermal capacity of water), air blowing from land over the water will become unstable in its lowest layers. Snow showers may develop in the now moist air over the lake. On reaching the opposite side of the lake, snowfall intensity increases because of local horizontal convergence. This is a result of the slight deceleration of the air caused by the increasing frictional effects of passage over land rather than the smooth water surface. Snowfall amounts decrease rapidly inland as surface cooling resumes, removing instability from the air.

Apparently subtle convergence effects can be a major impetus to cumulus development and shower formation globally, but nowhere are the weather effects as dramatic as here. The heaviest snowfalls tend to occur at the eastern ends of Lakes Erie and Ontario when surface (westerly) winds blow along the long axes of the lakes. Leathers and Ellis (1996) found that this synoptic type had increased in frequency since 1930, accounting for increased 'lake effect' snows – a snowfall hazard that is both severe and localized (Figure 5.25).

Figure 5.25 *Lake effect snows: NASA/SEAWiFS Project image of convective cloud forming over the Great Lakes on 4 February 2002 (courtesy of Goddard Space Flight Centre). The white tone to parts of the land surface represents distinct bands of snow cover falling over the previous week when severe ice-storms affected Oklahoma.*

(a)

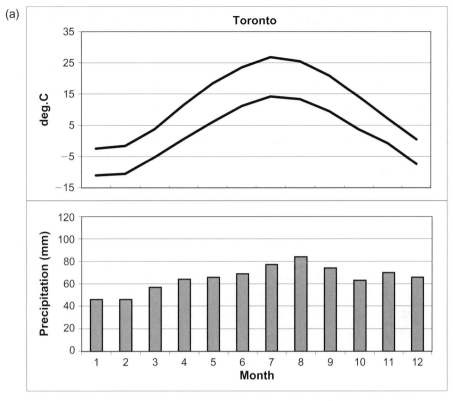

(b)

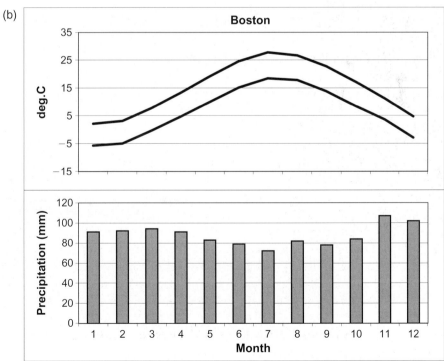

Figure 5.26 *Climate graphs for (a) Toronto and (b) Boston.*

The east of North America has a fairly uniform distribution of rain through the year; most places receive 1000–1200 mm, highest on the coast. This plentiful rainfall is due to the year-round presence of the jet stream and its enhancement by sharp temperature gradients between the eastern seaboard and the western North Atlantic. These thermal effects can invigorate depressions passing close to the eastern seaboard and heavy precipitation may develop at any time of year. The temperature contrast close to the east coast is maximized in cold spells in winter when the land is significantly colder than

---

Box 5.11 'The storm of the century' – 12–14 March 1993

A deep depression that tracked from the Gulf of Mexico to the northeast USA was responsible for record low air pressure and high daily snowfall accumulations. Snow depths exceeded 50 cm just northwest of the track (in the Polar air mass). On the warm side of the track close to the coast, the warmth of the sea surface encouraged low-level convection and thunderstorms.

Out of a total of 112 deaths, 26 people were killed in tornadoes, thunderstorms and flooding in Florida alone (Forbes et al., 1993). These multiple hazards originated from a squall-line that was probably enhanced by the near-record warmth around the Gulf of Mexico. Longer-term consequences included widespread beach erosion along the coast following the strong winds.

It is unusual for a single weather system to generate such a range of hazards over such a wide area; fatalities were reported from Cuba in the south to Quebec in the north. The storm resulted in total losses estimated at US$2.8 billion, of which US$1.75 billion was damage to insured property.

Box 5.12 The Canadian ice storm of 1998

Southeast Canada and the far northeast of the USA occasionally experience hazardous freezing rain events. Five major ice storms have hit the region since 1961 and the worst was probably that of January 1998 (Higushi et al., 2000).

The usual cause of freezing rain is the arrival of a warm front (carrying rain rather than snow) in an area that has just experienced sub-zero temperatures. Between 5 and 10 January 1998 two warm fronts moved very slowly over southeast Ontario and southern Quebec. Where warm air overlies cold, rain may fall onto a frozen ground surface, producing freezing rain: 70–100 mm of rainfall. A total of 39 people were killed in eastern Canada and the northeast USA (Dupigny-Giroux, 2000). These deaths were attributed to falling ice (and falling tree branches), hypothermia, carbon monoxide poisoning and house fires, the latter impacts arising from the extended power cuts caused by falling power lines. The effects of the storm were heightened by the fact that this rain was concentrated in densely populated areas around Montreal and Ottawa.

Higushi et al. (2000) have suggested that this weather situation is associated with the positive mode of the North Atlantic Oscillation in which a strong Azores High pushes upper westerly winds northeast from the Gulf of Mexico to southeast Canada.

the sea surface, producing a northwest to southeast temperature gradient. Autumn rainfall may be boosted by northward-tracking former hurricanes.

Large-scale temperature gradients encourage convergence uplift and deepening of low pressure systems tracking from southwest to northeast along the eastern seaboard. This coincides with the alignment of the jet stream emerging from the eastern Canadian trough; this can also locate the source area for this airstream over the Gulf of Mexico, a potent storm-generator.

As winter cooling of the land enhances the temperature contrast across the storm track, Atlantic coast 'Nor'easters' may develop, major depressions named after their direction of travel. These can give widespread blizzards that dislocate daily life over large areas, for example, 'the storm of the century' of March 1993 (Box 5.11). As mild airstreams become cooled moving north, freezing rain can become a widespread hazard (Box 5.12).

SUMMARY

Mid-latitude climates are characterized by tremendous variety. At the start of this chapter we highlighted latitudinal variations as a fundamental influence. This was related not just to temperature but also to the atmospheric circulation. Mid-latitude climates are found between the mid-latitude low pressure systems and sub-tropical high pressure; they are subject to westerly prevailing winds at all levels of the troposphere. The strength of this westerly circulation responds to the magnitude of latitudinal temperature contrasts – heightened contrasts resulting in a higher frequency of winter storms. Whether the atmospheric circulation will respond in the sme way to future global warming is an important but still rather uncertain issue.

The variety is greater still in the interiors of the land-masses of North America and Eurasia. In the more continental climate of North America the occurrence of tornadoes and severe thunderstorms is often associated with the contrasts between tropical and polar air, a different consequence of heightened latitudinal temperature contrasts compared with Europe. As in much of Asia, the influence of the westerly circulation is reduced by the effect of large mountain barriers, the enhanced role of meridional circulations boosting still further the continental character of climates here.

As a consequence of the arrangement of mountain barriers and sea areas, the maritime climates of Europe extend much further east than they do in North America. The equatorward decrease in summer rainfall is most clearly shown on the western sides of each continent; in the east, heavy summer rains of tropical origin extend polewards through the sub-tropics at the same latitude as the sub-tropical deserts further west. This also holds true in the southern hemisphere, although it could be said that mid-latitude climates are just as widespread in the southern hemisphere, but occur mostly over the sea! Overall, the influence of longitude is just as distinctive as that of latitude in both North America and Europe.

REFERENCES AND GENERAL READING

Black, A.R. and Bennett, A.M. 1995: Regional flooding in Strathclyde December 1994. In *Hydrological data UK: 1994 yearbook*. Wallingford: Institute of Hydrology.

Bowker, D. 2000: 'When the north wind blows' – the cold spell of April 1999. *Weather* 55, 315–20.

Burroughs, W.J. and Lynagh, N. 1999: *Maritime weather and climate*. London: Witherby.

Burt, S.D. 1980: Snowfall in Britain during winter, 1978/79. *Weather* 35, 288–301.

Burt, S.D. 1987: A new North Atlantic low pressure record. *Weather* 42, 53–56.

Burt, S.D. 1992: The exceptional hot spell of early August 1990 in the United Kingdom. *International Journal of Climatology* 12, 547–67.

Burt, S.D. 1993: Another new North Atlantic low pressure record. *Weather* 48, 98–103.

Burt, S.D. 1997: The Altnaharra minimum temperature of −27.2°C on 30 December 1995. *Weather* 52, 134–44.

Campins, J., Genovés, A., Jansà, A., Guijarro, J.A. and Ramis, C. 2000: A catalogue and a classification of surface cyclones for the Western Mediterranean. *International Journal of Climatology* 20, 969–84.

Dupigny-Giroux, L.-A. 2000: Impacts and consequences of the ice storm of 1998 for the North American north-east. *Weather* 55, 7–15.

Eden, G.P. 1995: *Weatherwise*. London: Macmillan.

Forbes, G.S., Blackall, R.M. and Taylor, P.L. 1993: 'Blizzard of the century' – the storm of 12–14 March 1993 over the eastern United States. *Meteorological Magazine* 122, 153–62.

Gray, M.E.B. and Marshall, C. 1998: Mesoscale convective systems over the UK, 1981–97. *Weather* 53, 388–96.

Higushi, K., Yuen, C.W. and Shabbar, A. 2000: Ice storm '98 in southcentral Canada and northeastern United States: a climatological perspective. *Theoretical and Applied Climatology* 66, 61–79.

Holt, M.A., Hardaker, P.J. and McLelland, G.P. 2001: A lightning climatology for Europe and the UK, 1990–99. *Weather* 56, 290–96.

Leathers, D.J. and Ellis, A.W. 1996: Synoptic mechanisms associated with snowfall increases to the lee of lakes Erie and Ontario. *International Journal of Climatology* 16, 1117–35.

Lott, J.N. 1994: The US summer of 1993. *Weather* 49, 370–83.

Marsh, T. 2001: The 2000/01 floods in the UK – a brief overview. *Weather* 56, 343–45.

Mayes, J. and Wheeler, D. 1997: The anatomy of regional climates in the British Isles. In Wheeler, D. and Mayes, J. (eds) *Regional climates of the British Isles*. London: Routledge.

McCallum, E. and Grahame, N.S. 1993: The Braer storm – 10 January 1993. *Weather* 48, 103–107.

Morton, O. 2000: The storm in the machine. *New Scientist* 31 January, 20–28.

Namias, J. 1987: Factors relating to the explosive North Atlantic cyclone of December 1986. *Weather* 42, 322–25.

Oliver, J.E. and Hidore, J.J. 2002: *Climatology, an atmospheric science*. Englewood Cliffs NJ: Prentice Hall.

Palutikof, J., Holt, T. and Skellern, A. 1997: Wind: resource & hazard. In Hulme, M. and Barrow, E. (eds) *The climates of the British Isles*. London: Routledge, 220–42.

Pearce, R., Lloyd, D. and McConnell, D. 2001: The post-Christmas 'French' storms of 1999. *Weather* 56, 81–91.

Pedgley, D.E. 1999: Shetland's wake in sea fog. *Weather* 54, 302–10.

Pedgley, D.E. 2000: A spot of winter warmth. *Weather* 55, 53–59.

Penarrocha, D., Estela, M.J. and Millan, M. 2002: Classification of daily rainfall patterns in a Mediterranean area with extreme intensity levels: the Valencia region. *International Journal of Climatology* 22, 663–76.

Perry, A.H. 2000a: The North Atlantic Oscillation: an enigmatic see-saw. *Progress in Physical Geography* 24, 289–94.

Perry, A.H. 2000b: Mediterranean climate. In King, R., Proudfoot, L. and Smith, B. (eds) *The Mediterranean: environment and society*. London: Arnold, 30–44.

Pytharoulis, I., Craig, G.C. and Ballard, S.P. 2000: The hurricane-like Mediterranean cyclone of January 1995. *Meteorological Applications* 7, 261–79.

Quaile, E.L. 2001: Back to basics: Föhn and Chinook winds. *Weather* 56, 141–45.

Rodwell, M.J., Rowell, D.P. and Folland, C.K. 1999: Oceanic forcing of wintertime North Atlantic Oscillation and European climate. *Nature* 398, 320–23.

Rowley-Gillard, L.H.G. 1996: Scotch on the rocks. *Weather* 51, 349–52.

Roy, M. 1997: Highland and Island Scotland. In Wheeler, D. and Mayes, J. (eds) *Regional climates of the British Isles*. London: Routledge.

Tout, D. and Wheeler, D. 1990: The early autumn storms of 1989 in eastern Spain. *Journal of Meteorology* 15, 238–48.

Ulbrich, U., Fink, A.H., Klawa, M. and Pinto, J.G. 2001: Three extreme storms over Europe in December 1999. *Weather* 56, 70–80.

Uriarte, A. 1980: Rainfall on the northern coast of the Iberian peninsula. *Journal of Meteorology* 49, 138–44.

Ustrnul, Z. 1992: The influence of föhn winds on air temperature and humidity in the Polish Carpathians. *Theoretical and Applied Climatology* 45, 43–47.

Wheeler, D. 1988: The Barcelona storm: 1st–5th October 1987. *Journal of Meteorology* 13, 79–86.

Wheeler, D. 1997: North East England. In Wheeler, D. and Mayes, J. (eds) *Regional climates of the British Isles*. London: Routledge.

Wheeler, D. 2001: Factors governing sunshine in south-west Iberia: a review of western Europe's sunniest region. *Weather* 56, 189–97.

Young, M.V. 1995: Severe thunderstorms over south-east England on 24 June 1994: a forecasting perspective. *Weather* 50, 250–56.

Young, M.V. and Grahame, N.S. 1999: Forecasting the Christmas Eve storm 1997. *Weather* 54, 382–91.

6

Tropical climates

Weather systems in the tropics are the product of the largest reservoir of energy in the atmosphere – the deep, warm and moist tropical troposphere that extends, on occasions, beyond 16–18 km above the Earth's surface. This warm air contains sufficient water vapour for condensation to provide an important additional energy source. Tropical climates are distinct from those of the mid-latitudes but it is increasingly realized that important interactions occur between the two climate systems. The best known of these is the influence of El Niño events on the weather and climate of many parts of the world. The aim of this chapter is to show how tropical climates are driven by thermal influences and how atmospheric and oceanic disturbances (e.g., El Niño events and tropical cyclones) shape the weather and climate of the tropics.

Figure 6.1 *A three-dimensional satellite view of Hurricane Mitch in the Caribbean, 27 October 1998 (courtesy of NASA).*

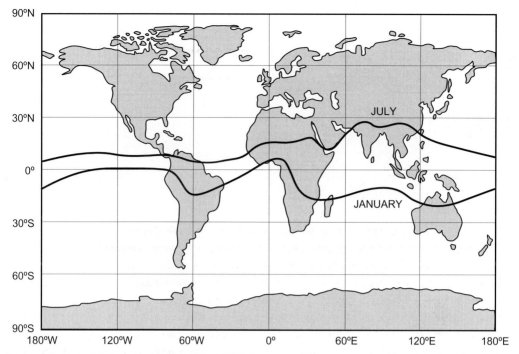

Figure 6.2 *Map of the average location of the ITCZ in January and July.*

6.1 IDENTIFYING AND DEFINING TROPICAL CLIMATES

The tropics are generally defined as those areas between the Tropics of Cancer and Capricorn (23.5°N/S), where the Sun is overhead for part of the year. The latitudinal range of the Sun can be regarded as the migrating meteorological equator, corresponding in theory with highest temperatures. In reality, the highest temperatures are found where cloud amounts are lowest and the heating of the overhead sun can be a powerful stimulus to cloud formation, an important negative feedback in the tropical atmosphere.

The focus of meteorological activity in tropical climates is the Inter-Tropical Convergence Zone (ITCZ), the zone of rising air, lifting in response to the heating of the overhead sun. This can be seen as the convergence zone of the atmospheric circulation of both hemispheres (Figure 6.2). The extent to which the ITCZ migrates is influenced by the blocking effect of high ground and by surface temperature variations, notably the heating of northern India in summer.

As the ITCZ is a zone of rising warm, often moist air, large volumes of water vapour are available for condensation. Although usually a zone of only shallow low pressure, it is identifiable on satellite imagery as an irregular band of convectional cloud clusters (Figure 6.3).

6.2 ATMOSPHERIC CIRCULATION, SURFACE TEMPERATURE AND CONVECTION

Surface temperature and the associated patterns of convection control the climate. The areas of highest surface temperatures tend to have the highest risk of convectional rainstorms. Heating here induces uplift – the rising branch of the Hadley cell (Section 2.3). This leads to the development of a shallow zone of low pressure along the ITCZ, encouraging the convergence of air. The rising air moves

Figure 6.3 *Meteosat thermal infrared image showing the position of the ITCZ close to the geographical equator at the autumn equinox, 2003 (06:00 h 24 September 2003). This image shows strong convection and well developed cloud systems around the ITCZ.*

polewards, especially on the western sides of the sub-tropical high pressure systems. This flow results in an important poleward heat transfer that helps to stabilize global temperatures.

Surface winds in these equatorial regions tend to be from the west. Areas lying under the influence of the ITCZ for most of the year (and therefore with rain throughout the year) are said to have an equatorial climate; the Amazon Basin in South America is a good example. This air cools adiabatically as it rises, enhancing the generation of convectional rainfall. In contrast, at around 35°N/S, the sub-tropics are characterized by the descending branch of the Hadley cell caused by the convergence of upper level air leaving the tropics and the mid-latitudes. This air is forced downwards, warming and drying adiabatically, creating the arid or semi-arid climates associated with the sub-tropical anticyclones.

The zones of tropical wetness and sub-tropical dryness migrate with the overhead Sun. As a result, areas between the equatorial belt and the sub-tropical deserts experience summer rains and a winter dry season; i.e., the tropical wet/dry climate. This is the zone of the tropical easterly winds – the northeast Trade winds of the northern hemisphere and the southeast Trades of the southern hemisphere. The season of summer rains is shortest towards the sub-tropics where the ITCZ is only present for a short period in mid-summer. This progression from wet to dry is most clearly evident in Africa and southern Asia where the ITCZ migrates over the longest distances.

UPPER WINDS AND JETS
The tropics and sub-tropics contain two important jet streams in the upper troposphere. The sub-tropical jet is found at a height of 12 km around the 200 mbar level. On satellite imagery, it is often seen over

North Africa as a thick band of cirrus extending northeastwards, particularly in the early months of the year. Clear air turbulence, an important aviation hazard, is often experienced around its edges.

The tropical easterly jet is found between India and the equator around the 150 mbar level. It reaches the strength of a jet stream in July and August around 10°N, developing as a result of the reversal of the north–south pressure gradient between India and the equator at this time of year.

6.3 DISTURBANCES IN TROPICAL CLIMATES (I): TROPICAL CYCLONES

Tropical cyclones are the most obvious expression of atmospheric energy in tropical climates. These weather systems are called hurricanes in the Atlantic, cyclones in the south Pacific and Indian Ocean, and typhoons over the northwest Pacific: these are regional names for the same meteorological phenomenon, tropical revolving low pressure systems. A sequence of atmospheric disturbances can be identified.

- **Squall lines** are linear cloud systems hundreds of kilometres long and 10–25 km wide. They can be identified on satellite imagery and sometimes show a convex leading (front) edge, although extensive trailing anvils of cumulonimbus cloud may hide lower cloud features. They are an important source of convectional rainfall.
- **Easterly waves** develop from squall lines and are low pressure waves usually having a southwest to northeast orientation in the northern hemisphere (northwest to southeast in the southern hemisphere). They develop with the squall line within 15°N/S of the equator and are blown west by the Trade winds; this motion is an important aspect of the future development of the system as the depth of the unstable moist layer of air increases in this direction in both the Pacific and the Atlantic; the tropical temperature inversion that places a cap on this convection rises as the waves travel west.
- **Tropical depressions** are low pressure centres having mean wind speeds up to 34 knots. Approximately one in four easterly waves develop into tropical depressions.
- **Tropical storms** develop from tropical depressions, with mean wind speeds between 34 and 64 knots. Approximately one in ten easterly waves develop into tropical storms.
- **Tropical cyclones** have mean wind speeds of more than 64 knots and are usually more compact than mid-latitude depressions: they typically have diameters less than 600 km. The main distinguishing feature is a warm centre, culminating in the 'eye', a 20 km diameter core of warm, gently subsiding air. The intensity and destructive power of tropical cyclones varies hugely and is classified according to the Saffir–Simpson scale (Table 6.1).

Table 6.1 Mean wind speeds and air pressure associated with the Saffir–Simpson scale

Category	Mean wind speed			Typical air pressure (mbar)
	$m s^{-1}$	knots	mph	
1	33–43	64–83	74–95	>980
2	43–50	83–97	96–110	979–965
3	50–59	97–114	111–130	964–945
4	59–69	114–134	131–155	944–920
5	70+	136+	>155	<920

The impact of the severest systems is increased by the gustiness of the winds. This is especially true for coastal regions that usually experience the severest conditions: maximum gusts may increase 50% above mean speeds here, though inland a figure of 30% is typical.

ENERGY SOURCES

The principal source of energy for a tropical cyclone is a surface temperature of at least 27°C. This heat must also be sufficiently widespread through the ocean and the atmosphere to withstand the mixing that occurs as a result of the rising wind speeds. In addition to the warm core that initiates the convection, several other factors are essential to sustain the evolution of a tropical storm into a tropical cyclone:

- condensation of rising air (releasing the latent heat of condensation);
- vorticity – this provides the energy to force rotation. This is a function of latitude and is not present within 5° of the equator so tropical cyclones tend to occur (independently) either side of the equator but not close to it;
- low wind shear – a consistent wind speed and direction through the troposphere allow the rising cloud systems to retain their structure as they ascend. This is a factor that disables the formation of tropical cyclones in the Mediterranean Sea and other extra-tropical seas where sea temperatures may sometimes reach 27°C.

The air within a tropical cyclone is lifted by vigorous convection rather than the frontal convergence found in mid-latitude depressions. Consequently, they are low pressure systems without fronts and this helps to explain their circular appearance. The relatively smooth appearance on satellite imagery reveals merely the texture of the top of the cloud system and obscures the convection cells (containing thunderstorms) underneath. The smooth cap is actually a mass of ice crystals – in the form of cirrostratus cloud – that forms because of the cooling that accompanies uplift to 15 km or more (Aguado and Burt, 2002).

IMPACT AND EFFECTS

The impact of tropical cyclones can be measured in human, economic and physical terms. The ranking of individual tropical cyclones will vary according to the differing vulnerability of the areas affected. About 15% of the world's population are said to be at risk from the effects of tropical cyclones, the human impact being severest in the eastern rather than western hemisphere. This is illustrated by the worst tropical cyclone documented in modern times; the East Pakistan cyclone of 1970 (in what is now Bangladesh) was responsible for the deaths of at least 300 000 people. Despite advances in warning systems and shelters, another cyclone in 1991 caused the deaths of 70 000 in the same region.

Only about 20% of tropical cyclones reach category 3 or more on the Saffir–Simpson scale but these are responsible for around 80% of tropical cyclone damage in the USA. Hurricane Andrew in 1992 (Figure 6.4) caused more damage (over US$25 billion) in under four hours than any other single climatic hazard. This was a consequence of the track passing over southern Florida, including the southern suburbs of Miami, a relatively short land track. Gusts exceeding 150 knots generated a 4.5 m storm surge. In contrast to Bangladesh, only 15 fatalities were attributed directly to the storm (Rappaport, 1994).

Hurricane Gilbert in September 1988 was physically the most intense Atlantic hurricane on record: air pressure dropped to a record 885 mbar with sustained winds of more than 150 knots. It had the second highest economic impact (US$10 billion), behind only Hurricane Andrew. Gilbert tracked along the length of Jamaica, remaining as a category 5 storm for 31 hours (Figure 6.5). By 16 September it had reached northern Mexico and was downgraded to a tropical storm the following day. However, it continued to give copious amounts of rain and severe flooding claimed the lives of 200 people.

Figure 6.4 *Hurricane Andrew over the Gulf of Mexico, 25 August 1992. This image was taken after it had passed over southern Florida (courtesy of NASA).*

Figure 6.5 *NOAA GEOS West and GOES East image of Hurricane Gilbert (12:00 h 12 September 1988). This was taken after it had crossed Jamaica as it approached the Yucatan peninsula (courtesy of NOAA).*

The annual frequency of tropical cyclones in the Atlantic has remained fairly constant over the past half-century. However, peaks in hurricane incidence arouse concerns over the potential effects of global warming. In 1995 11 hurricanes were reported in the Atlantic, double the long-term average. Global warming might lead to a wider geographical distribution if the threshold sea surface temperature of 27°C occurs over a larger area. In autumn 1999, North Carolina was affected by three major hurricanes, the first such occurrence since 1954. These storms had a serious impact on an agricultural economy ill-adapted to such tropical extremes. A total of 2.8 million chickens and turkeys, 880 cattle and 30 000 pigs were killed; 22 dams failed and severe pollution of coastal waters and estuaries ensued (Leetma and Crowder, 1999). Hurricane Isabel followed a similar track in September 2003.

The most destructive western hemisphere tropical cyclone of the last 200 years was Hurricane Mitch in October 1998 (Figure 6.1). It caused severe flooding and landslides in Honduras and northern Nicaragua (Hellin and Haigh, 1999), caused severe damage to the infrastructure and resulted in the deaths of more than 11 000 people. Wind speeds averaged over one-minute intervals reached 156 knots.

The highest rainfall accumulations caused by tropical cyclones have been recorded on La Réunion Island in the Indian Ocean. Tropical cyclone Hyacinthe gave 3240 mm of rain in 72 hours in January 1980 and as much as 5678 mm over ten days. The frequency and intensity of tropical cyclones over much of the Indian Ocean peaks during La Niñas. Severe flooding in Mozambique in early February 2000 was aggravated by the arrival of tropical cyclone Eline on 21 February. This brought down power lines and severely affected the response to the worsening flood hazard.

In February 2003, four tropical cyclones were present simultaneously over the Indian Ocean (Figure 6.6).

Figure 6.6 *NASA image of a series of four tropical cyclones over the Indian Ocean in February 2003. A MODIS Terra image (courtesy of NASA).*

PREDICTION, FREQUENCY AND FORECASTING

Recent advances in forecasting achieved through the wider use of satellite images (Elsberry and Velden, 2003) have made it easier to forecast the track of a tropical cyclone. Most of the steering influence depends on air pressure and upper winds in the surrounding area. Forecasts of track in the late 1990s made 48 hours in advance were nearly as accurate as those made just 24 hours in advance in 1970.

The intensity of tropical cyclones is more challenging to predict than the track. Kerry Emanuel (1999) presented a skilful array of retrospective simulations of intensity based on the absolute temperature difference between the warm sea surface and the top of the troposphere. Model simulations quantify the degree to which warm surface waters mix with cooler water below – an important influence on cyclone intensity because rising wind speeds cool the surface by raising cool, deep water towards the surface. Other factors are the state of the surrounding atmosphere and the internal workings of the storm.

There is considerable scientific uncertainty regarding the response of tropical cyclones to global warming. In theory a rise in sea surface temperatures from 27°C to 28°C would be expected to raise the frequency of storms (and to increase their geographical extent). The latter factor, though, is likely to be limited by the presence of vertical wind sheer as latitude increases.

6.4 DISTURBANCES IN TROPICAL CLIMATES (2): EL NIÑO EVENTS

El Niño events affect weather patterns around the world and demonstrate the complex (and little understood) interrelationships that exist within the ocean–atmosphere system, and between temperature, air pressure and winds. The phenomenon is an outstanding case of a teleconnection, a long distance association in the global climate system: while flooding creates havoc in parts of South America and Australasia, India and Africa suffer severe drought conditions.

Strictly speaking, the name El Niño applies to just a small part of the disturbance. To avoid confusion, the definitions of the main terms encountered in this topic can now be reviewed:

- **El Niño** is the name given to the periodic warming of the tropical southeast Pacific ocean off the coast of Ecuador and Peru resulting from a change in ocean currents. It was often noted by fishermen around Christmas, hence the name, the Spanish for Christ child. It induces low pressure and heavy rain in areas that are arid at other times. It is associated with a temporary interruption to the normal upwelling of cool, nutrient-rich water that supports a rich variety of sea-life off the coast of Peru and Ecuador. El Niño events therefore have a disastrous effect on the local fishing industry and the marine ecology (Glantz, 1993).
- **La Niña** describes the opposite phase in this cycle when the warm water is displaced by unusually cold currents: the name means 'little girl' in Spanish.
- The **Southern Oscillation** is a measure of the inverse relationship that exists between the air pressures recorded over northern Australia and the tropical central Pacific (e.g., around Tahiti). It is an index of the redistribution of air pressure between the western and central/eastern tropical Pacific, a useful indicator of the transition between El Niño and La Niña states.

El Niño and La Niña events were well known as local peculiarities in the climate system of South America for centuries before the Southern Oscillation was identified. The discovery of the Southern Oscillation was linked with one of the consequences of an El Niño – a disastrous drought in 1877 that killed possibly over 20 million people in southern Asia and parts of Africa (Anderson, 1999). This led to the realization that El Niño events were part of a larger-scale climatic oscillation. An important link was made by the British meteorologists Sir Gilbert Walker and Sir Norman Lockyer who examined relationships between Indian rainfall and climatic datasets of different areas. Their research revealed a 'see-saw' pattern (negative correlation) of air pressure and rainfall between the tropical Indian and Pacific Oceans. In doing so they identified the most important teleconnection of the global climate system.

EFFECTS OF EL NIÑO AND LA NIÑA EVENTS

These events have a significant impact on climate and society at local, regional and global scales. For example, at a relatively local level, an entire marine ecosystem can be disrupted by changes in nutrient supply (e.g., upwelling of deep ocean water maintains a supply of zooplankton and phytoplankton that helps to sustain the natural fertility of the Peru current). The arrival of warm, nutrient-poor water in an El Niño reduces local populations of anchoveta fish with consequent knock-on effects on bird life. At a regional level normally arid areas experience flooding and, at the global scale, much of the entire climate system is perturbed.

At the regional scale there is a direct effect upon air temperature through the interaction between the sea surface and the atmosphere. This in turn influences rainfall by redistributing air pressure and areas experiencing convection. Weather conditions are then further modified by changes in wind strength and direction caused by the redistribution of air pressure. During El Niños areas of strongest convection and high rainfall are found over northwestern South America, the central equatorial Pacific, the central western equatorial Indian Ocean and off the coast of equatorial West Africa. In a La Niña event, enhanced convection and rainfall shift to Australasia, northeastern South America and parts of equatorial Africa, the Indian sub-continent and northern China (Allan *et al.*, 1996).

Statistically significant associations have been found between the ENSO cycle and rainfall in many parts of the world. These represent interesting manifestations of teleconnections but, for much of the mid-latitudes, there are many more important influences on rainfall variations. An exception is in North America where several distinct regional associations with ENSO occur. For example, in an El Niño event, the Gulf States and Great Lakes region tend to be wetter than average.

EL NIÑO CIRCULATION OF THE ATMOSPHERE AND OCEAN

In low latitudes a layer of warm, buoyant water occupies the shallowest layers with colder, denser water at greater depths. The boundary between these layers is called the **thermocline**. In an average year, the anticlockwise winds circulating around the sub-tropical high pressure in the south Pacific move the warm surface water away from the coast of Peru and Ecuador allowing cold water to rise from below the surface (Figure 6.7). This is called the 'Walker circulation', in honour of Sir Gilbert Walker. This brings the thermocline near to the surface and keeps the sea surface temperature close to 19°C rather than in the mid-20s, which might be expected at this latitude. The cold ocean surface discourages convection, helping to maintain dry conditions along the coast, and also encourages the continuation of high pressure.

The anticlockwise circulation around the high pressure centre feeds a much larger scale circulation – the Trade winds which blow from east to west across the tropical Pacific, both sides of the equator. These are driven by the pressure gradient between high pressure off South America and low pressure around Indonesia that forms over very high sea surface temperatures. The winds drag the water surface west, the depth of warm water increasing westwards as it warms on its long journey towards Indonesia. As a consequence of this warming, the level of the ocean can be raised by 20 to 30 cm. The warm air rises readily over Indonesia and a cellular pattern of airflow then occurs, with air returning eastwards in the upper troposphere (Figure 6.7). Consequently, the Trade winds weaken, sometimes reversing to a westerly direction. This further inhibits upwelling since the mechanism for removing surface water has stopped. The thermocline is subsequently forced downwards as masses of warm water accumulate in the eastern tropical Pacific.

An El Niño can start from one of two events. The first is the inflow of a warm, equatorial current along the coast of Peru and Ecuador. This caps the upwelling of cold water and reduces the height of the thermocline. The second occurs when a rising pool of warm water in the west bursts eastwards across the tropical Pacific.

What follows is an excellent demonstration of atmosphere–ocean coupling. As the sea warms, the air above warms and air pressure falls, weakening the temperature and air pressure gradients (contrasts) across the Pacific region. Consequently, the Trade winds weaken, sometimes reversing to a westerly

(a)

El Niño Conditions

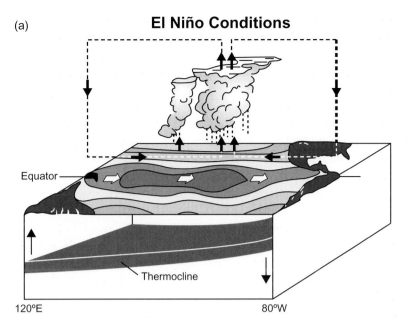

Equator

Thermocline

120°E   80°W

(b)

La Niña Conditions

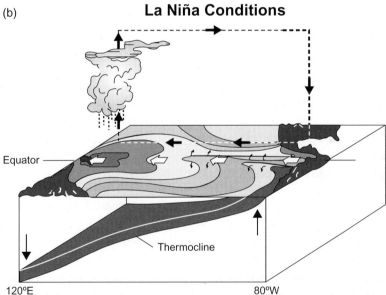

Equator

Thermocline

120°E   80°W

Figure 6.7 *Diagrammatic representation of (a) El Niño and (b) La Niña circulations (courtesy of the NOAA TAO El Niño Theme Page: http://www.pmel.noaa.gov/tao/elnino/nino-home.html – last accessed 19 November 2003).*

direction, and this further inhibits upwelling since the mechanism for removing surface water has stopped. The thermocline is subsequently forced downwards as masses of warm water accumulate in the eastern tropical Pacific. This warmth induces convection and the formation of low pressure systems. The climatic consequence is one of heavy rain and flooding along a usually dry coastline.

La Niña events comprise the reverse anomalies of air pressure, surface temperature, wind direction and weather to El Niño. It can be a hazardous situation in its own right. Drought affects parts of the South American west coast and the risk of severe floods in Australia, Indonesia and India increases.

Box 6.1 Burning biomass

Vegetation fires perturb local and global biogeochemical systems, disrupt economies and eco-systems and affect human health. The exact impact of these carbon-loaded fluxes on the Earth's energy system and weather remains uncertain.

Fires release carbon-rich aerosols and trace gases that disrupt short- and long-wave radiances (scatter and absorb energy), affect clouds and modify rainfall patterns. Nucleation is activation by the aerosols but rainfall is suppressed because the small droplets do not coalesce to form larger droplets and so remain suspended in cloud (Nober et al., 2003): a connection between fires, rainfall suppression and local drought events is suggested.

Many parts of the world are experiencing serious economic, social and environmental problems associated with unprecedented episodes of intensive and geographically extensive fires. Satellite data revealed that in two years (post 2000) $500\,000\,km^2$ of Australia was burnt. The highest aerosol signal detected in space is regularly found near the tropics (e.g., Indonesia, Amazon Basin, sub-Saharan Africa). The fires, rarely 'wild-fires', are triggered by a whole gamut of complex and interrelated climate and human factors. Imagery is invaluable for locating, tracking and quantifying smoke aerosols, and identifying source hot spots providing timely information that assists in the process of fire management.

In September 1997, satellite images showed that smog blanketed over 4 million km^2 of southeast Asia; extreme pollution lasted for many weeks, causing severe health problems and a serious hazard to air, sea and land traffic. Although, at the time, unusually dry weather associated with an El Niño event promoted conditions conducive to biomass burning (Li and Ramanathan, 2002), spatial analysis (in a geographical information system) of fire hot spots using remotely sensed thermal imagery demonstrated that burning was not associated with natural forest fires but inappropriate land-use practices, mostly associated with commercial operations (Stolle and Tomich, 1999).

In April 2001 NOAA scientists reported a huge pall of smoke over parts of Canada and the USA that had originated from China (Brown, 2001). At source, the plume had been up to 3 km thick, covered 10 million km^2 and reputably blocked 15% of the sunlight. Fires were attributed to human activities, linked principally to Chinese land-use policies and practices (Hong Yang and Xiubin Li, 2000). Climate factors linked to an El Niño event and shifts in monsoon rains made a serious on-going problem a catastrophic event.

Globally, smoke plumes are regularly detected over forests mostly as a result of non-accidental fires. Images from the Space Shuttle's final flight reveal that the normally dense cloud cover over the Amazon Forest is locally reduced by biomass burning. Dark particles absorb energy and warm the surrounding air inhibiting condensation: the local and global impact of this phenomenon on weather and climate is unknown.

Anderson (1999) has discussed how the Indian Ocean differs from both the Pacific and the Atlantic. The east is warmer than the west because of westerly surface winds. Deviations from this pattern are called the Indian Ocean dipole; late in 1997 low pressure and relatively high temperatures in the west gave severe flooding to parts of east Africa; the water level in Lake Victoria rose rapidly by 2 m. At the same time Indonesia suffered drought (Box 6.1). Although this might appear to be consistent with an

El Niño (a strong El Niño was prevailing at the time), this pattern can also occur for other reasons (Saji *et al.*, 1999).

EL NIÑO EVENTS IN HISTORY

The negative impact of El Niño events has been recorded by Peruvian fishermen for centuries but, until recently, there was no long-term chronological record. Dan Schrag (cited in Pearce, 1999) has analysed growth records of coral as a record of El Niño activity, confirming a 100 000 year chronology. Details of the strength and regularity of the events are not yet known through this period. This is an important aspect of setting recent records in perspective because a change in the behaviour of El Niño has been noted in recent decades. Since about 1976, they have become more frequent; the return period has decreased from an average of 6 years to only 3.5 years and some recent events have been unusually persistent, notably that of 1990–95.

This raises the intriguing question of whether the periodicity (cyclic pattern) has been disturbed and, if so, by what factor? Some scientists have discussed the possibility of global warming being the driving force for this change. Indeed, Kevin Trenberth (cited in Pearce, 1999) has identified a rise in winter sea surface temperatures in the tropics around 1976 and a 10 m reduction in the level of the thermocline has been studied by Guilderson and Schrag (1998). These changes are consistent with the results of some global warming models but their causes remain poorly understood.

6.5   THE CLIMATES OF AFRICA

DISTRIBUTION OF CLIMATE ZONES

The African continent is unique in occupying both hemispheres and shows an approximate symmetry of climatic zones about the equator (Figure 6.3). It provides a good opportunity to observe the distribution of global climate zones.

- **Equatorial climates** occur around the equator as far east as Lake Victoria. This climate is characterized by rain at all seasons and a very small annual range in temperature. Locations near the equator have rainfall maxima in spring and autumn, when the ITCZ is overhead, and minima around January and July when it is more distant, as exemplified by Lagos (Figure 6.8).
- **Tropical wet/dry climates** are found on either side of equatorial climates, extending north to the Sahel (15°W to 15°E; 15–20°N) and south to Zambia and northern Mozambique (e.g., Khartoum in Figure 6.8). This is a zone of dry winters and wet summers, most of the rain being related to the location of the ITCZ. The drier 'Trade wind' climates of southern Mozambique, Zimbabwe and northeastern South Africa have a longer dry season. The relevance of the Trade winds lies in the exposure to these moisture-bearing winds on the eastern side of Africa whereas the west coast of southern Africa at this latitude lies in a desert and semi-desert environment.
- **The sub-tropical deserts** of the Sahara (25–30°N) and the Kalahari (20–25°S) are found beyond the seasonal range of the ITCZ, under the descending air of the sub-tropical high pressure systems. The dryness of the western side of southern Africa is associated not just with the sub-tropical high pressure but with the stability of the eastern South Atlantic, cooled by the Benguela current. Sea breezes here bring fog rather than showers.
- **Mediterranean type climates** are found on the poleward sides of the Sahara and Kalahari; from Morocco to Tunisia and in the Cape Province (as shown by Cape Town in Figure 6.8).

VERTICAL VARIATIONS IN TEMPERATURE AND INSTABILITY

The Trade wind inversion height increases towards the ITCZ allowing a greater depth of warm, moist air within which convection can develop. The deepest cumulus cloud develops some distance from

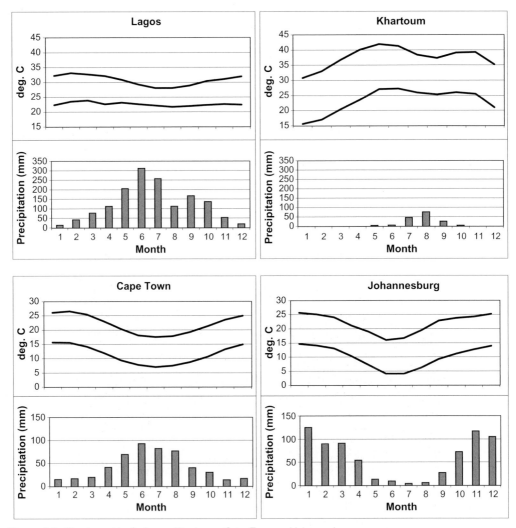

Figure 6.8 *Climate graphs for Lagos, Khartoum, Cape Town and Johannesburg.*

the zone of surface convergence; the weather therefore lags behind the surface location of the ITCZ. Temperature inversions also characterize the thermal structure of the air around the cooler oceans off Namibia and Angola and around the Canary Islands.

Satellite images of the sub-tropical oceans reveal a great variety of instability patterns in cloud. Figure 6.9 shows the distribution of cloud and dust off the coast of northwest Africa. Plumes of Saharan dust appear as light grey (see Box 6.2). The coastal zone is clear of cloud when the prevailing Trade winds blow dry air off the Sahara. However, banks of shallow stratocumulus often form in the relatively stable but moist air further offshore around the Canary Islands. These are draped over the more mountainous islands, especially between May and September when their onshore movement is aided by sea breezes. Above 1500 m on Tenerife, Gran Canaria and La Palma near-desert conditions prevail – this is above the Trade wind inversion for much of the year, a feature that caps the development of low-level cloud sheets (Buckle, 1996).

Box 6.2 Globalization of dust aerosols

Dust affects climate and weather, human health, local and global economies, the environment, and the Earth's energy balance. Once high in the atmosphere, huge quantities of particulate matter become dispersed and unconstrained, its global distribution controlled by macro-scale circulation systems.

Source-to-sink tracking of dust fluxes is possible from space because the small particles (0.04–3.00 µm) vigorously scatter visible and NIR (short-wave) radiances and observations are frequent and at an appropriate scale. In VIS images, plumes are easily detected over dark (non-reflective) ocean surfaces but are less obvious over land (see Figure 6.9). Dust plumes can be distinguished over arid/semi-arid regions in thermal IR images because they have a significant diurnal impact on IR emissions (Legrand et al., 2001).

Figure 6.9 *Visible image of cloud and dust off the coast of northwest Africa (courtesy of NASA).*

Analysis of archived satellite data indicates that dust-loading of the atmosphere has increased since the 1960s (Mahowald and Kiehl, 2003). Scientists believe that this is contributing to perturbations in the Earth's energy balance and may possibly be affecting rainfall patterns: the precise effect on climate is uncertain (Tanre et al., 2003).

The principal source region for desert dust is North Africa, particularly eastern Mauritania, western Mali and southern Algeria (Goudie and Middleton, 2001). Prevailing winds, controlled by the seasonal pattern of pressure systems, determine the trajectories of dust plumes originating from the Saharan region. For example, the eastern Mediterranean is affected during low pressure conditions (Michaelides *et al.*, 1999).

Dust interacts with clouds promoting cloud-cover changes that affect incoming and outgoing radiances. For example, in less than two decades, low cloud cover has increased over the tropical Atlantic but high cloud cover has decreased (Mahowald and Kiehl, 2003). High dust concentrations promote nucleation in low clouds but the droplets remain small so that rainfall is suppressed: this may contribute to anomalous drought conditions leading to more dust emissions. However, the ice crystals in high convective clouds attract moisture, enlarge and fall. This promotes rain and reduces cloud cover at high altitudes.

Saharan dust regularly reaches the Caribbean, Florida and beyond: over a period of three months an estimated 13 million tonnes of dust reached the northeastern regions of the Amazon Basin. The dust is always a mixture of mineral particles and organic matter and both are the focus of environmental and health-related research. Significant quantities of bacteria and fungi spores get a free ride across the Atlantic and, although most are harmless to humans, some bacteria are pathogens and represent a possible health hazard (Griffin *et al.*, 2001). Desert dust is rich in iron (an essential nutrient) and plays a major role in oceanic productivity. The amount of biologically usable nitrogen available in the marine ecosystem is increased by complex biogeochemical processes, possibly responsible for triggering eutrophication and toxic algal blooms (red tides), and contributing to the decline of coral in the Caribbean (Griffin *et al.*, 2001).

Several times a year the UK is affected by red dust originating from the Sahara; occasionally exceptionally high levels of particulates are recorded and air quality is affected. In March 2000, high dust levels were at first thought to have originated from a volcanic event in Iceland but satellite imagery and field measurements indicated that the most likely origin was the Sahara (Ryall *et al.*, 2002).

Severe dust storms are generated regularly in the Gobi Desert, mostly during the spring. In April 1998, two particularly intense dust storms occurred that reduced direct solar radiation by 30–40%, increased the albedo of the Pacific Ocean by 10–20% and polluted the west coast of North America to such an extent that the aerosols presented a health hazard to vulnerable people. The plumes were monitored by satellite and there was an exceptional exchange of information between an *ad hoc*, international Web-based virtual community of researchers.

ATMOSPHERIC INSTABILITY IN EQUATORIAL AFRICA

The development of cumulus over equatorial Africa fluctuates not just on a daily timescale in response to diurnal temperature changes but also over longer periods. Meso-scale convective systems (Box 5.7) may form, sometimes coalescing into cloud areas extending for more than 500 km. These contain several 'hot towers' of rising air and thunderstorms. At other times cloud is far more broken and the ITCZ is difficult to identify.

Variations in cloud amount are related to feedback processes linked to surface temperature (Charney, 1975). When temperatures increase under clear skies, air near the ground becomes unstable, rises and is forced to cool. If it is cooled to dewpoint temperature, cumulus cloud develops. This triggers two negative feedbacks: cloud cover reduces the surface receipt of solar radiation, leading to a loss of sensible (direct) heat energy, and latent heat is lost from the surface when moisture is evaporated. Both of these processes lead to lower surface temperatures, weaker convection and a return to clear skies and sunshine.

Dust can be regarded as both a consequence of and a cause of drought in the sub-tropics, the latter arising from the cooling effect of low-level dust. Furthermore, desertification often induces a rise in albedo through loss of vegetation that cools the surface and further discourages rainfall.

Equatorial East Africa (east of Lake Victoria) is surprisingly dry and does not experience the all-year rains typical of true equatorial climates. This is due to reduced instability associated with the Northeast Monsoon and the tropical easterly jet. Between December and March, the monsoon wind dries as it passes over the cool Somali current in the Indian Ocean so less moisture is available for cloud development. The jet stream, which achieves maximum strength between July and August around 10°N, induces upper troposphere convergence (and hence air divergence below), that inhibits the development of summer rains – especially over Eritrea, and sometimes further west towards the Sahel region.

In coastal areas of West Africa, surface air convergence can be a significant local stimulus for rainfall. The wettest areas occur where prevailing onshore winds trigger convergence as their velocity is reduced on landfall. The driest areas are where the prevailing wind is off-shore and is therefore subject to horizontal divergence arising from reduced surface friction over the sea.

EFFECTS OF EL NIÑO EVENTS IN AFRICA

Parts of Africa, especially the Sahel and southern Africa, experience drought triggered by El Niño events. In Sudan, dry years, associated with reduced summer rainfall, correlate inversely with ENSO events ($r = -0.34$) and high sea surface temperatures in the Indian Ocean ($r = -0.57$) (Osman and Shemseldin, 2002). In contrast, parts of East Africa receive more rain during El Niño events. In Kenya, the October–January rains are enhanced, but the March–June rains remain unaffected (Amissah-Arthur et al., 2002). Overall, although the El Niño 'signal' changes from place to place, variability in rainfall tends to increase.

6.6 MONSOON CLIMATES OF SOUTH ASIA

Sea surface temperatures around the Indian Ocean are high throughout the year. They peak in spring when the area exceeding 29°C covers most of the Indian Ocean north of 10°S. Average temperatures over the other seasons are around 28°C (Webster et al., 2002). It is within this environment that the ITCZ reaches its most northerly point, around 30°N over Nepal and southern China in July (Figure 6.2). This energy is augmented by the intense heating of the Indian sub-continent in spring and summer, giving rise to a distinctive temperature distribution and the most dramatic of monsoon climates.

THE SEASONAL PROGRESSION OF THE SOUTH ASIAN MONSOON CLIMATE

Few places in the world can rival the consistency of weather changes between the seasons found in the Indian sub-continent (O'Hare, 1997). Intense heating in spring and summer induces the decrease in atmospheric pressure that accompanies the southwest monsoon rains. In winter, the region is influenced by the development of the Siberian high pressure, although it rarely suffers the low temperatures experienced in central Asia.

Spring is a 'weatherless' season; the whole region often appears completely free of cloud on satellite imagery. The surface absorbs increasing amounts of solar radiation and the dryness of the land speeds the conversion of this energy to heat. Temperatures over the land reach 35°C relatively early in spring, the high sea surface temperatures failing to moderate the heating, unlike in most mid-latitude environments.

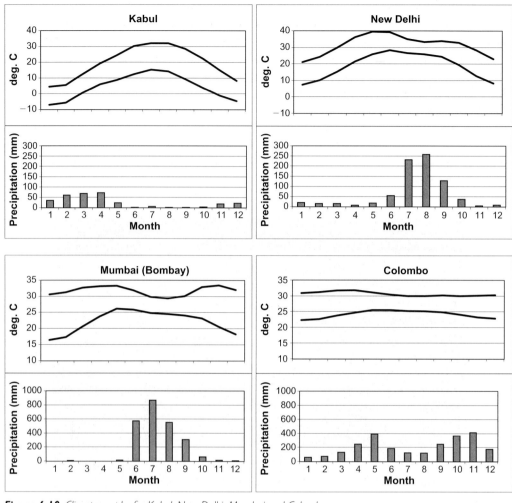

Figure 6.10 *Climate graphs for Kabul, New Delhi, Mumbai and Colombo.*

Strong spring insolation up to the start of the monsoon rains accounts for the peak in maximum day temperatures in early summer, resulting in an asymmetrical temperature curve (Figure 6.10).

A 'heat low' develops in response to the heating over South Asia, the prelude to the arrival of the monsoon rains. This pushes the sub-tropical high pressure system (the Arabian High) west over the Arabian Sea. Meanwhile, the 'meteorological equator' is now located over central India. This induces a cross-equatorial jet, a long distance airflow that moves northwest across the Indian Ocean. This is deflected to the northeast and moves towards India, driven by the surface pressure gradient towards the heat low. This circulation is boosted by an interesting interaction between tropical and mid-latitude airstreams; the mid-latitude westerly circulation flips to the northern side of the Himalayan plateau in early summer.

The warm airflow carries the energy and moisture that is responsible for initiating the heavy monsoon rains that usually reach southern India by June, and the Ganges Valley by early July. Rain is restricted to the Nepalese foothills when short periods of drier weather occur – locally known as the 'break monsoons'.

By mid-October, the rains have shifted from the northwest and affect central and southern India: October to December is the wettest period in southern India and Sri Lanka. Spring and autumn are also wet seasons in Sri Lanka, associated with the Yala and Maha rains that coincide with the overhead passage of the ITCZ (see Colombo, Figure 6.10). The autumn rains are associated with a wind shift to the northeast and a consequent exposure to the depressions and cyclones developing over the Bay of Bengal (46% of cyclones occur here in October and November). In winter, the rest of India is dry and relatively cold as the Northeast Monsoon brings continental air to the region.

RAINFALL EXTREMES IN SOUTH ASIA

The high rainfall experienced in South Asia during the Southwest Monsoon is comprised of relatively widespread monsoon rains, sometimes associated with tropical cyclones or monsoon depressions and smaller convective cells that initiate showers. As in other parts of the world, these showers are characterized by greater rainfall intensities and tend to occur when skies are relatively clear, away from the most active phases of the monsoon rains (Figure 6.11).

Figure 6.11 *Multi-spectral AVHRR image of tropical cyclone 04B (the Orissa cyclone), 17 October 1999 (courtesy of NASA).*

Box 6.3 Impact of El Niño on Indian harvests

Indian grain harvests have increased dramatically since the mid-twentieth century in response to the adoption of food-grain varieties that are more responsive to fertilizer. Total yields of rice, wheat and pulses quadrupled between 1950 and 2000 from around 50 million tonnes per annum to around 200 million tonnes (Selvaraju, 2003). Instead of being a steady response to these non-climatic factors, actual annual yields varied according to the strength of monsoon rainfall; yields are inversely related to El Niño strength ($r = -0.50$), significant at the 1% level. Grain production fell in 12 out of 13 El Niño events in the late twentieth century, resulting in an average economic cost of US\$773 million. In contrast, 10 out of 13 La Niña years saw an increase in yield, which provided an average economic gain of US\$437 million. The average monsoon rainfall anomaly in El Niño years was −14% whereas in La Niña years it was +9%.

Rainfall data should be interpreted cautiously in such a volatile and variable climate. Average annual rainfall across India varies from around 50 mm in the Thar desert in the northwest to 11 000 mm at Cherrapunji, on the Khasi Hills of Assam in northeast India. The highest yearly rainfall yet recorded globally occurred here in 1974: 24 550 mm and 26 470 mm was recorded within 12 months in 1860–61. The wettest day on record here yielded 1563 mm, twice the annual rainfall of lowland Britain (Beresford and O'Hare, 1999). The Khasi Hills are aligned east–west, well placed to experience strong orographic uplift of the Bay of Bengal branch of the Southwest Monsoon. Uplift is also encouraged by upper tropospheric easterly winds in summer.

VARIATIONS IN THE MONSOON RAINS

Meehl (1994) has suggested a link between snowfall over the Himalayas, the Indian monsoon and sea surface temperatures (SST) around India over a two-year sequence. Light snowfall in winter on the Himalayas (with high SST around India) precedes a strong summer monsoon (associated with low SST). In the following winter, snowfall increases in the Himalayas (accompanied by low SST around India) resulting in a weak summer monsoon (and high SST) the following year. This relationship between the monsoon rains and local sea temperature anomalies may at first seem surprising. It occurs because the very warm ocean in winter triggers a stronger than average monsoon, associated with windy conditions and higher evaporation rates – both having a cooling effect on the sea surface (Webster et al., 2002). Sea surface anomalies tend to persist long enough to influence the monsoon of the following year – in this case, encouraged by a cooling of the Indian Ocean.

There is some evidence that year-to-year variability in the monsoon rains is limited by a negative feedback between the tropical ocean and atmosphere (Webster et al., 2002). In summer, the Southwest Monsoon circulation in the lower atmosphere provides a significant northward transport of heat. This coexists with a southward flow of heat in the Indian Ocean. In winter, the roles of atmosphere and ocean are reversed (initiated by the southward flow of surface winds). These circulations are enhanced in El Niño events as sea surface temperatures are reduced, helping to offset slightly the effect of the El Niño itself.

SUMMARY

Tropical climates evolve from the thermal conditions of the coupled ocean–atmosphere system of the low latitudes. Disturbances to average or 'normal' conditions are an important source of climatic variability and uncertainty in many parts of the world. The El Niño–La Niña cycle and tropical cyclones are two sources of major (and interlinked) disturbances, both of which interact with mid-latitude climates. A more general uncertainty for local communities is the generation of rainfall, usually an end-result of convection, a fact that highlights the fundamental importance of thermal influences and vertical motion on tropical climates.

REFERENCES AND GENERAL READING

Agvado, E. and Burt, J.E. 1999: *Understanding weather and climate*. Upper Saddle River, New Jersey: Prentice Hall.

Allan, R., Lindesay, J. and Parker, D. 1996: *El Niño Southern Oscillation and climate variability*. Collingwood, Australia: CSIRO.

Amissah-Arthur, A., Jagtai, S. and Rosenzweig, C. 2002: Spatio-temporal effects of El Niño events on rainfall and maize yield in Kenya. *International Journal of Climatology* 22 (15), 1849–60.

Anderson, D. 1999: Extremes in the Indian Ocean. *Nature* 401, 337–39.

Beresford, A.K.C. and O'Hare, G. 1999: A comparison of two heavy rainfall events in India: Bombay, 24 July 1989, and Cherrapunji, 12 June 1997. *Weather* 54, 34–43.

Bigg, G.R. 1996: *The oceans and climate.* Cambridge: Cambridge University Press.

Brown, L.R. 2001: Dust bowl threatening China's future. http://www.earth-policy.org/Alerts/Alert13.htm. See also http://www.epa.gov/air/airtrends/asian_dust4.pdf.

Buckle, C. 1996: *Weather and climate in Africa.* Harlow: Addison Wesley Longman.

Charney, J.G. 1975: Dynamics of deserts and droughts in the Sahel. *Quarterly Journal of the Royal Meteorological Society* 101, 193–202.

Diaz, H.F., Hoerling, M.P. and Eischeid, J.K. 2001: ENSO variability, teleconnections and climate change. *International Journal of Climatology* 21, 1845–62.

Elsberry, R.L. and Velden, C. 2003: A survey of tropical cyclone forecast centres – uses and needs of satellite data. *World Meteorological Organisation Bulletin* 52, 258–64.

Emanuel, K. 1999: Thermodynamic control of hurricane intensity. *Nature* 401, 665–69.

Glantz, M.H. 1993: *Currents of change: El Niño's impact on climate and society.* Cambridge: Cambridge University Press.

Goudie, A.S. and Middleton, N.J. 2001: Saharan dust storms: nature and consequences. *Earth-Science Reviews* 56, 179–204.

Griffin, D.W., Garrison, V.H., Herman, J.R. and Shinn, E.A. 2001: African desert dust in the Caribbean atmosphere: microbiology and public health. *Aerobiologia* 17, 203–13.

Guilderson, T.P. and Schrag, D.P. 1998: Abrupt shift in subsurface temperatures in the tropical Pacific associated with changes in El Niño. *Science* 281, 240–46.

Hellin, J. and Haigh, M.J. 1999: Rainfall in Honduras during Hurricane Mitch. *Weather* 54, 350–59.

Hong Yang and Xiubin Li 2000: Cultivated land and food supply in China. *Land Use Policy* 17 (2), 73–88.

Lau, K.-M. and Busalacchi, A.J. 1993: El-Niño – Southern Oscillation: a view from space. In Gurney, R.J., Foster, J.L. and Parkinson, C.L. (eds) *Atlas of satellite observations related to global change.* Cambridge: Cambridge University Press.

Leetma, A. and Crowder, L.B. 1999: The climate of 1999: La Niña, drought and hurricanes. *US Global change research programme seminar programme.* Washington DC: http://www.usgcrp.gov (last accessed 19 November 2003).

Legrand, M., Plana-fattori, A. and N'doume, C. 2001: Satellite detection of dust using the infrared imagery of Meteosat 1. Infrared difference dust index. *Journal of Geophysical Research – Atmospheres* 106 (D16), 18251–74.

Li, F. and Ramanathan, V. 2002: Winter to summer monsoon variation of aerosol optical depth over the tropical Indian Ocean. *Journal of Geophysical Research* 107 (D16).

Mahowald, N. and Kiehl, L. 2003: Mineral aerosols and cloud interactions. *Geophysical Research Letters* 30 (9), 109.

Meehl, G.A. 1994: Influence of land surface in the Asian summer monsoon: external conditions versus internal feedback. *Journal of Climate* 7, 1033–49.

Michaelides, S., Evripidou, C.P. and Kallos, G. 1999: Monitoring and predicting Saharan desert dust events in the eastern Mediterranean. *Weather* 54, 359–65.

Nober, F.J., Graf, H.F. and Rosenfeld, D. 2003: Sensivity of the global circulation to the suppression of precipitation by anthropogenic aerosols. *Global and Planetary Change* 37 (1–2), 57–80.

O'Hare, G. 1997: The Indian monsoon, part 2: the rains. *Geography* 82, 335–52.

Osman, Y.Z. and Shemseldin, A.Y. 2002: Qualitative rainfall prediction models for central and southern Sudan using El Niño – Southern Oscillation and Indian Ocean sea surface temperature indices. *International Journal of Climatology* 22, 1861–78.

Pearce, F. 1999: Weather warning. *New Scientist* 164, 36–39.

Rappaport, E.N. 1994: Hurricane Andrew. *Weather* 49, 51–61.

Ryall, D.B., Derwent, R.G., Manning, A.J., Redington, A.L., Corden, J., Millongton, W., Simmonds, P.G., O'Doherty, S., Carslaw, N. and Fuller, G.W. 2002: The origin of high particulate concentrations over the United Kingdom, March 2002. *Atmospheric Environment* 36, 1363–78.

Saji, N.H., Goswanmi, B.N., Vinayachandran, P.N. and Yamagata, T. 1999: A dipole mode in the tropical Indian Ocean. *Nature* 401, 360–63.

Selvaraju, R. 2003: Impact of El Niño-Southern Oscillation on Indian food-grain production. *International Journal of Climatology* 23, 187–206.

Stolle, F. and Tomich, T.P. 1999: The 1997–1998 fire event in Indonesia. *Nature and Resources* 35, 22–30.

Tanre, D., Haywood, J., Pelon, J., Leon, J.F., Chatenet, B., Formenti, P., Francis, P., Goloub, P., Highwood, E.J. and Myhre, G. 2003: Measurement and modelling of the Saharan dust radioactive impact: Overview of the Saharan Dust Experiment (SHADE). *Journal of Geophysical Research – Atmospheres* 108, 8574.

Timmermann, A., Oberhuber, J., Bodier, A., Esch, M., Latif, M. and Roeckner, E. 1999: Increased El Niño frequency in a climate model forced by future greenhouse warming. *Nature* 398, 694–97.

Webster, P.J., Clark, C., Cherikova, G., Fasullo, J., Han, W., Loschnigg, J. and Sahami, K. 2002: The monsoon as a self-regulating coupled ocean–atmosphere system. In Pearce (ed.) *Meteorology at the millennium.* Oxford: Academic Press, 198–219.

Index